Electromagnetic Field Measurements in the Near Field

Electromagnetic Field Measurements in the Near Field

Hubert Trzaska

Noble Publishing Corporation
Atlanta

Library of Congress Cataloging-in-Publication Data

Trzaska, Hubert
    Electromagnetic field measurements in the near field / Hubert Trzaska
      p. cm.
    Includes index.
    ISBN 10884932-10-X
      1. Electromagnetic fields--Measurement

    QC665.E4 T79 2001
    530.14'1'0287--dc21

00-068702

Copyright © 2001 by Noble Publishing Corporation.

All rights reserved. No part of this book may be reproduced in any form by any means without prior written permission of the publisher.

Printed in the United States of America

ISBN 1-884932-10-X

Contents

Preface ix

Chapter 1   Introduction 1
    1.1 Bibliography 10

Chapter 2   **The Principles of Near-Field EMF Measurements** 13
    2.1 An EMF generated by a system of currents 13
    2.2 The far field and the near field 17
    2.3 EMF from simple radiating structures 22
    2.4 Bibliography 28

Chapter 3   **EMF Measurement Methods** 29
    3.1 E, H and S measurement 30
    3.2 Temperature rise measurements 36
    3.3 Current measurements 41
    3.4 Bibliography 45

Chapter 4   **Electric Field Measurement** 47
    4.1 Field averaging by a measuring antenna 49
    4.2 Influence of fields from beyond a probe measuring band 52
    4.3 Mutual interaction of hte measuring antenna and the field source 64
    4.4 Changes for the probe's directional pattern 69
    4.5 The E-field probe comparison 77
    4.6 Comments and conclusions 82
    4.7 Bibliography 85

Chapter 5	**Magnetic Field Measurement** 87
	5.1 Measuring antenna size 87
	5.2 Frequency response of the magnetic field probe ... 90
	5.3 Directional pattern alternations 94
	5.4 Accuracy of measurement versus distance of the antenna to the source of radiation 98
	5.5 The magnetic field probe with a loop working above its self-resonant frequency 104
	5.6 Comments and conclusions 109
	5.7 Bibliography 112
Chapter 6	**Power Density Measurement** 113
	6.1 Power density measurement methods 113
	6.2 Power density measurement using the antenna effect 124
	6.3 Conclusions and comments 132
	6.4 Bibliography 136
Chapter 7	**Directional Pattern Synthesis** 137
	7.1 A probe composed of linearly dependent elements 138
	7.2 Spherical radiation pattern of an E/H probe 142
	7.3 A probe composed of three mutually perpendicular dipoles 146
	7.4 Comments and conclusions 153
	7.5 Bibliography 156
Chapter 8	**Other Factors Limiting Measurement Accuracy** 159
	8.1 Thermal stability of a field meter 159
	8.2 The dynamic characteristics of the detector 167
	8.3 Measured field deformations 171
	8.4 Susceptibility of the meter to external EMF 172
	8.5 Resonant phenomena 175
	8.6 Bibliography 179

Chapter 9	**Photonic EMF Measurements**	**181**
	9.1 The photonic EMF probe	186
	9.2 Frequency response of the probe	189
	9.3 Sensitivity of the photonic probe	195
	9.4 Linearity of the detector	199
	9.5 Synthesis of the spherical directional pattern	201
	9.6 The future meter	203
	9.7 Bibliography	206
Chapter 10	**Final Comments**	**209**
Index		**217**

Preface

The proliferation of electronic devices has dramatically increased the number of sources of electromagnetic fields (EMF). Public awareness and professional concern have been combined in government regulations and voluntary standards that place limits on the intensity of these fields over various frequency ranges and in different environmental situations. This book covers the methods for measuring EMF to verify compliance with these regulations and standards, and also to provide accurate data in the research required for the development of new standards.

These regulations and standards address two concerns. The first is interference, when fields radiated from one device affect the operation of others. In most cases, this is merely an inconvenience or annoyance, but if the interference affects navigation or emergency communications systems, there can be unfortunate consequences. The second concern is biological. The heating effects of high concentrations of electromagnetic energy are well-known, while research is continuing into the long-term effects of EMF exposure below that which causes measurable heating. This latter situation is the subject of headline news coverage and much public conjecture. The research required to ascertain the precise biological effects of EMF must include accurate field measurements.

Hubert Trzaska is an international expert in EMF measurement techniques. The near field measurement theory and techniques described in this book are essential for any engineer or scientist who works in this area. The author's well-reasoned commentary adds valuable insight into the practical aspects of EMF measuring equipment, its accuracy and its proper use.

Gary A. Breed
President, Noble Publishing Corp.

1

Introduction

The degradation of the natural electromagnetic environment, which has been led to the edge of an ecological disaster and sometimes beyond, is the forgotten price that must be paid for our inconsiderate enthusiasm for "industrial revolution." As a result, we are reaching the situation where spending for the protection of the environment must sometimes exceed the investment in the systems causing the degradation.

The development of contemporary civilization is associated with consumption of more and more quantities of energy in forms that are applicable in technology, science, medicine and in our households. One form of energy with a rapidly growing role in everyday life is the energy of RF currents and fields. In some applications the energy is a final product (telecommunication or radiolocation), while in the others it is an intermediate form, designed to be transferred, for instance, into heat. In both cases there can be intentional or unintentional radiation of part of that RF energy, and, as a result, contamination of the whole environment and interference over a wide frequency range.

The natural electromagnetic environment can be understood as fields naturally occurring in the biosphere: the electric field (E), the magnetic field (H) and the electromagnetic field (EMF). Into this natural environment came global wireless communication systems and power systems based upon the alternating current. Recently, we have observed a trend to return to wire (fiberoptic) data transmission and to global satellite systems to meet the qualitative and quantitative necessities of telecommunication. As a result, the development of large-scale dominating wireless communication systems has been halted. Simultaneously, however, there has been an explosion in the popularity of wireless communication systems for local communication (wireless phones, cellular phones, radiotelephones, CB, remote control devices) as well as EMF generating devices, especially in the household (microwave ovens, dielectric and inductive heating, video display monitors). These systems are causing the whole global population to exist an electromagnetic environment for which the adjective "natural" was lost 50 to 70 years ago.

Among the distinctive features of the natural EMF environment degradation, as compared to the other forms of the environmental pollution, are these:

- It is a unique realm where the pollution is caused intentionally (telecommunication)
- Its pollution is largest, acting immediately and on a global scale
- The exposure of the people working in the vicinity of the biggest power sources (telecommunication) is much below those near the medium power sources (industry, science, medicine, household) and even low power (mobile communication)
- It is the single area where there is a theoretical possibility to eliminate the pollution completely, without any remains.

Investigations of the biological activity of currents, electric and magnetic fields reach as far back as ancient times and were intensified with the technology of EMF generation [1, 2]. These investigations have included applications in medical diagnostics and therapy as well as hazards created by the interaction of these factors with the human organism. It is worth noting here to analogies between the hazard created by an artificial EMF and that caused by contact with the natural fields [3, 4], as well as the separation of the biosphere from extraterrestrial fields [5] (with an exception of two "windows" at frequencies where the atmosphere is transparent to radio waves).

The electromagnetic field, apart from a narrow frequency band and within limited amplitude range, is not detectable by organoleptic methods. Thus, EMF detection and all work and investigation related to this the field requires the use of tools. Moreover, EMF is not directly measurable and it is necessary to transfer it to an another quantity that we are able to measure (voltage, heat).

EMF measurement in the far-field (Fraunhofer zone) is one of the less accurate measurements of physical quantities. Hazardous exposure to EMF requires field measurements in the neighborhood of primary and secondary field sources as well as fields disturbed by the presence of materials and the transmisson media. Our attention must be focused on the near-field (Fresnel region). The near-field conditions cause farther degradation of the near-field EMF measurements' accuracy as compare to those in the far-field. These difficulties raise doubts about the measuring equipment and its users, and brings frustration to its designers. Although not considered in this work, the "Achilles heel" here is the accuracy of the EMF standards. A standardized device can not be more accurate than the standard used for the procedure. Remarkably, at the present time the accuracy of a "good" EMF standard does not exceed ±5%.

This book is devoted to the specific problems of EMF measurements in the near- field and to the analysis of the main factors limiting the measurement accuracy, especially in the near-field. It is focused upon the measurements included with the regulations for labor safety and general public protection against unwanted exposure to EMF. These issues represent the involvement of the author, however, almost identical metrological problems exist in widely understood electromagnetic compatibility (EMC) methods. The analyses presented here make it possible to estimate the importance and the role of various factors involved in specific conditions of a measurement, as well as evaluation of available (offered on the market) meters (and their manufacturers).

4 ELECTROMAGNETIC FIELD MEASUREMENTS IN THE NEAR FIELD

The measurements for surveying or monitoring require the use of quantities that are relatively simple to measure, with meters that fulfill conditions of reliability and accuracy, although measurements in difficult field or industrial conditions are sometime performed. Contrary to laboratory investigations or research, where arbitrarily selected quantities could be the subject of the interest and measurement (if they are correctly selected, applied in conditions fulfilling the basic methodological requirements and while repeatability of the measurement is assured), because of practical reasons, we will limit ourselves only to the quantities which indications form the name HESTIA — the goddess of fireside and the natural environment [6]. The quantities, derived quantities and several constants useful in the further considerations are shown in Table 1-1.

Quantity	Symbol	International unit (SI)
Magnetic field strength	H	amperes per meter [A/m]
Electric field strength	E	volts per meter [V/m]
Power density	S	watts per sq. meter [W/m^2]
Temperature	T	kelvins [K]
Current intensity	I	amperes [A]
Magnetic flux density	B	tesla [T] = 10^4 gauss[G]
Current density	J	amperes per sq. meter [A/m^2]
Specific Absorption	SA	joules per kilogram [J/kg]
Specific Absorption Rate	SAR	watts per sq. meter [W/kg]
Conductivity	σ	siemens per meter [S/m]
Permittivity	ε	farads per meter [F/m]
Permittivity of vacuum	ε_o	ε_o = 8.854 10^{-12} F/m
Permeability	μ	henrys per meter [H/m]
Permeability of vacuum	μ_o	μ_o = 12,566 10^{-7} [H/m]

Table 1-1. *Quantities representing EMF and their units.*

In free space and in non-magnetic media the magnetic flux density (B) is equal in value to the magnetic field intensity (H). Some meters (especially those devoted for magnetostatic field and VLF alternating fields) are calibrated in B-units. In order to make the conversion of the units easier, Table 1-2 presents their relationship within the range essential for practical applications.

A/m	796	80	8	0.8	80 mA/m
gauss [G]	10	1	0,1	10 mG	1 mG
tesla [T]	1 mT	0.1 mT	10 µT	1 µT	0.1 µT

Table 1-2. *Corresponding values of H field units in non-magnetic medium.*

The parameters of field strength meters are, in the area of applications discussed here, especially in Poland, precisely given by standards [8, 9, 10, 11]. However, they are only partly related to the specific types of required near-field measurements. From a considered point of view, there are usually unnecessary parameters defined in the standards; and sometimes conditions required by the standards and have definitions that are not understandable. On the other hand, the most essential parameters are presented (if any) without any comments that would make it possible to analyze the measurement conditions and the domain in which the meter may be successfully applied (within its limits of accuracy). The latter is easy to understand as it was not the subject of the documents. In order to better introduce present and future metrological needs, Table 3 below lists the EMF exposure limits in accordance with standards in force in Poland, as well as some fraction of the newest proposals worked out by national and international competent teams. Because of the availability of these documents and their current modifications, only uncontrolled environment limits are below presented to illustrate the metrological needs.

6 ELECTROMAGNETIC FIELD MEASUREMENTS IN THE NEAR FIELD

Frequency Range	E [V/m]	H [A/m]	S [W/m^2]	I [nA/m^2]
Static field	16,000	8,000		100
50 Hz	10,000	80		
0.001 – 0.1 MHz	100	10		
>0.1 – 10 MHz	20	2		
>10 – 300 MHz	7			
>0.3 – 300 GHz			0.1	

Table 1-3. *Exposure limits in protection zones under Polish regulations [12].*

Frequency Range	E [V/m]	H [A/m]	B [µT]	S [W/m^2]
>0 – 1 Hz	10,000	3.2 x 10^4	4 x 10^4	
>1 – 8 Hz	10,000	3.2 x 10^4/f^2	4 x 10^4/f^2	
>8 – 25 Hz	10,000	4,000/f	5,000/f	
>0.025 – 2.874 kHz	250/f	4/f	5/f	
>2.874 – 5.5 kHz	87	4/f	5/f	
>5.5 – 100 kHz	87	0.73	0.91	
>0,1 –1 MHz	87	0.23/f$^{-1/2}$		
>1 – 10 MHz	87/f$^{-1/2}$	0.23/f$^{-1/2}$		
>10 – 400 MHz	27.5	0.073		2
>400 – 2000 MHz	1,375 f$^{1/2}$	0.0037 f$^{1/2}$		f/200
>2 – 300 GHz	61	0.16		10

Table 1-4. *Permissible exposure levels in accordance to the IRPA proposals [7].*

Type of risk	Frequency Range	Conduction Current [mA]
Professional	1 Hz – 2.5 kHz	1.0
	2.5 kHz – 100 kHz	0.4 × f
	100 kHz – 100 MHz	40
General public	1 Hz – 2.5 kHz	0.5
	2.5 kHz – 100 kHz	0.2 × f
	100 kHz – 100 MHz	20

Table 1-5. *Permissible current intensity in a hand or in a foot [7].*

In Tables 1-4 and 1-5, f (frequency) is in the units indicated in the column titled "frequency range."

In the United States two proposals have been worked out recently [12, 13]. Both of them, with regards to the permissible exposure, are similar to the US national standard ANSI/IEEE C95.1-1992. Somewhat different levels of exposure are given in proposals presented by ACGIH and are related to the controlled environment [14].

8 ELECTROMAGNETIC FIELD MEASUREMENTS IN THE NEAR FIELD

Frequency Range	E [V/m]	H [A/m]	S [W/m²] (PE) (PH)	T_{AV} [H] (E) (S)
3 kHz – 100 kHz	614	163	10^3 10^7	6 6
100 kHz – 1.34 MHz	614	16.3/f	10^3 $10^5/f^2$	6 6
1.34 MHz – 3.0 MHz	823.8/f	16.3/f	$1800/f^2$ $10^5/f^2$	$f^2/0.3$ 6
3.0 MHz – 30 MHz	823.8/f	16.3/f	$1800/f^2$ $10^5/f^2$	30 6
30 MHz – 100 MHz	27.5	$58.3/f^{1.668}$	2 $9.4 \times 10^6/f^{3.336}$	30 $0.0636 f^{1.337}$
100 MHz – 300 MHz	27.5	0.0729	2	30 30
300 MHz – 3 GHz			f/150	30
3 GHz – 15 GHz			f/150	90000/f
15 GHz – 300 GHz			100	$616000/f^{1.2}$

Table 1-6. *Exposures permitted by the American proposals [12, 13].*

Type of risk	Frequency Range [MHz]	Maximal current of both feet [mA]	Maximal current of a foot [mA]	Conduction current [mA]
Professional	0.003 – 0.1	2000f	1000f	1000f
	0.1 – 100	200	100	100
General Public	0.003 – 0.1	900f	450f	450f
	0.1 – 100	90	45	45

Table 1-7. *Permissible currents induced by the EMF [12, 13].*

In Tables 1-6 and 1-7, f = frequency in MHz, T_{AV} = average time in minutes

The above cited proposals of IRPA standards, as well as the American ones, are based upon detailed studies of the biomedical and physical issues. Especially well-founded are proposals of the IRPA [7], and the progress of the studies is currently published [15, 16, 17]. Although *"the time between formulation of the proposals to their implementation may be as long as from the Acropolis construction to the proposals' formulation,"* [18] nevertheless, just now they may be useful for showing the direction of further metrological needs.

The author, as an electronic engineer, has never reserved himself any right to suggest what should be the exposure limits and always warns against a mechanistic approach to the bioelectromagnetic problems, but it seems that just such an approach was dominant while the present standards were worked out. It is impossible to believe that the bioeffects are so precisely known that it was possible to propose standards with an accuracy to the third decimal point, not to mention the possibility of field strength measurement with such an accuracy! Apart from the controversial question on the levels presented in the tables, their citation is not to present the author's support to these levels but only as an introductory estimation of the EMF strength measurement range or, rather, to establish the upper limits of measured fields since the lower ones could be below the noise level of the most sensitive meters. Ultimately, the exposure limits should be proposed by biologists and medical doctors, and physicists and engineers should have an auxiliary, although indispensable, role. The limits prepared in such a way could be a bit less precise but they surely will be much more humanitarian. A trend in this direction has already been demonstrated [19, 20].

1.1 Bibliography

1. The Decree of the Minister of Environment Protection, Natural Resources and Forestry, Sept. 11, 1998 relating to the detailed rules of protection against electromagnetic radiation harmful to humans and the environment, exposure levels permissible in the environment and valide requirements when radiation surveying is performed (in Polish), Dz.U. No. 107/98, pos. 676.
2. S. M. Michaelson, M. Grandolfo, A. Rindi, *Historical Development of the Study of the Effects of ELF Fields. In: Biological Effects and Dosimetry of Static and ELF Electromagnetic Fields*, pp. 1–14. Plenum Press, 1985.
3. R. A. Waver, *The Electromagnetic Environment and the Circadian Rhythms of Human Subjects*, Ibid, pp.477–524.
4. A. S. Presman, *Electromagnetic Field and the Life* (in Russian). Moscow 1968.
5. J. Holownia, *Natural Sciences, Geopathical Zones and Radiesthesy*, Technical Univ. of Wroclaw 1993.
6. H. Trzaska, "Power Density as a Standardized Quantity," *COST 244 WG Meeting*, Athens 1995, pp.111–118.
7. *Guidelines on limits of exposure to time-varying electric and magnetic fields and to radiofrequency electromagnetic fields (1 Hz – 300 GHz)*. Draft, IRPA/INIRC 1994.
8. PN-77/T-06581 – *Labour protection against EMF within frequency range 0.1 – 300 MHz. EMF meters.* (Polish standard).
9. PN-89/T-06580/02, *Labour protection against EMF within range 1 – 100 kHz. EMF meters.* (Polish standard).
10. *Measuring equipment for electromagnetic quantities*, Prepared by IEC TC 85 WG11.
11. *Radio transmitting equipment. Measurement of exposure to radiofrequency electro-magnetic field - field strength in the frequency range 100 kHz to 1 GHz,* IEC SC12C.
12. The Decree of the Council of Ministry of Nov.5, 1980 related to the detailed rules of the protection against electromagnetic nonionizing radiation harmful for humans and for the environment (in Polish), Dz.U. No. 25/80, pos.101).

13. "Guidelines for evaluating the environmental effects of radiofrequency radiation," ET Docket No. 93–62
14. *NATO Standardization Agreement 2345: Control and evaluation of personnel exposure to radio frequency fields.*
15. "Threshold Limit Values for Physical Agents in the Work Environment," Adopted by ACGIH with Intended Changes for 1994–1995.
16. "Electromagnetic Fields (300 Hz to 300 GHz)," Environmental Health Criteria 137, WHO, Geneva 1993.
17. ICNIRP Guidelines. "Guidelines on limits of exposure to static magnetic fields," *Health Physics*, nr 1/1994, pp.100–106.
18. B. Kunsch, "The new European Pre-Standard ENV 50166 – Human exposure to electromagnetic field," *COST 244 Working Group Meeting*, Athens 1995, pp.48–58.
19. H. Trzaska, "What about frequency independent standards?" *Proc. 4th EBEA Congress*, p. 121–122, Zagreb, 1998.
20. B. Eicher, "Bioelectromagnetics: The Gap Between Scientific Knowledge and Public Perception," *Proc. 1999 Int'l EMC Symp.*, Zurich, pp. 71–76.

2

The Principles of Near-Field EMF Measurements

In order to illustrate the problems related to near-field EMF measurements, we will derive formulas which can then be applied to a discussion of the factors limiting field measurement accuracy. We will also compile material that is necessary for readers to perform their own analyses of the problems considered in their work, as well as other problems related to near-field measurements under various conditions and circumstances. The essential information for practical metrology is presented in this chapter, including a brief summary of the near-field properties as well as the basic equations and formulas related to fields generated by simple radiation sources.

2.1. An EMF generated by a system of currents

Let's assume that within the volume V there exists a system of arbitrarily oriented electric and magnetic currents **J** and ***J** respectively. The volume V is surrounded by an infinitely large, homogeneous, isotropic, linear, lossless medium. Its electrical properties are described by the permeability ε and the permittivity μ; apart from it there is no discontinuity of electrical parameters on

14 ELECTROMAGNETIC FIELD MEASUREMENTS IN THE NEAR FIELD

the boundary surface. The maximal linear size of the volume V, in an arbitrary cross section, is D (Figure 2.1).

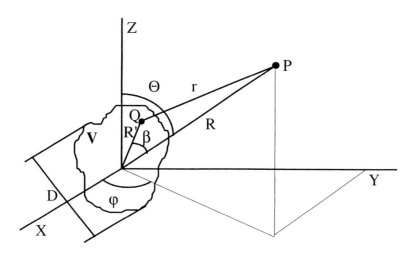

Figure 2.1 *EMF in point P generated by currents in volume V.*

Solving Maxwell's equations for the above-formulated boundary conditions, for a monochromatic harmonic oscillation of angular frequency ω, we find the electric field strength vector E and the magnetic vector H given at an arbitrary point of observation P (R, Θ, φ), that is situated outside the volume V [1]:

$$\mathbf{E} = \nabla \times \nabla \times \mathbf{\Pi} + j\omega\mu \nabla \times {}^*\mathbf{\Pi} \qquad (2.1)$$

$$\mathbf{H} = \nabla \times \nabla \times {}^*\mathbf{\Pi} - j\omega\varepsilon \nabla \times \mathbf{\Pi} \qquad (2.2)$$

where $\mathbf{\Pi}$ and ${}^*\mathbf{\Pi}$ = the electric and magnetic Hertzian vectors:

$$\Pi = \frac{1}{j4\pi\omega\varepsilon} \int_V \mathbf{J} \frac{\exp(-jkr)}{r} dV \qquad (2.3)$$

$${}^*\Pi = \frac{1}{j4\pi\omega\mu} \int_V {}^*\mathbf{J} \frac{\exp(-jkr)}{r} dV \qquad (2.4)$$

where k = the propagation constant:

$$k = \omega\sqrt{\varepsilon\mu} \qquad (2.5)$$

r = the distance from the observation point P to an integration point Q (R', Θ', φ'). In vector notation, it takes the form:

$$\mathbf{r} = \mathbf{R} - \mathbf{R}'$$

while its magnitude is:

$$r = \sqrt{R^2 + R'^2 - 2RR'\cos\beta} \qquad (2.6)$$

where

β = an angle between R and R',
R = the distance from the observation point to the center of the coordinate system
R' = the distance from the point of integration to the center of the coordinate system

For R'<R, r may be presented with the use of a series expansion (Equation 2.7):

$$r = R\left[1 - \frac{R'}{R}\cos\beta + \frac{R'^2}{2R}(1 - \cos^2\beta) + \frac{R'^3}{2R^3}\cos\beta(1 - \cos^2\beta) - \cdots\right] \qquad (2.7)$$

If R>>D (where D is the maximal size of an arbitrary cross section of the volume V), it is possible to assume that r is parallel to R, so $r \approx R - R' \cos \beta$. Then:

$$\Pi_\infty = \frac{\exp(-jkR)}{j4\pi\omega\varepsilon R} \int_V \mathbf{J} \exp(-jkR' \cos \beta) \, dV \qquad (2.8)$$

$${}^*\Pi_\infty = \frac{\exp(-jkR)}{j4\pi\omega\mu R} \int_V {}^*\mathbf{J} \exp(-jkR' \cos \beta) \, dV \qquad (2.9)$$

The index ∞ in the formulas indicates that they are valid for R >> D. In the case the spatial components of E and H are given by:

$$E_{\theta\infty} = -jk(Z\Pi_{\theta\infty} + {}^*\Pi_{\varphi\infty}) \qquad (2.10)$$

$$E_{\varphi\infty} = -jk(Z\Pi_{\varphi\infty} - {}^*\Pi_{\theta\infty}) \qquad (2.11)$$

$$E_{R\infty} = H_{R\infty} = 0 \qquad (2.12)$$

$$H_{\varphi\infty} = \frac{E_{\theta\infty}}{Z} \qquad (2.13)$$

$$H_{\theta\infty} = \frac{-E_{\varphi\infty}}{Z} \qquad (2.14)$$

where:

$\Pi_{\theta\infty}$, $\Pi_{\varphi\infty}$, ${}^*\Pi_{\theta\infty}$ and ${}^*\Pi_{\varphi\infty}$ = the spatial components of vector Π_∞ and ${}^*\Pi_\infty$

Z = wave impedance of the medium,

$$Z = \sqrt{\frac{\mu}{\varepsilon}} = Z_0 \sqrt{\frac{\mu_r}{\varepsilon_r}} \qquad (2.15)$$

Z_0 = intrinsic impedance of free space:

$$Z_0 = \sqrt{\frac{\mu_0}{\varepsilon_0}}$$

Formulas (2.10 – 2.14) allow us to find the far-field EMF components of an arbitrary system of currents in volume V. The field may be characterized as follows:

- The EMF in the far-field is the transverse one (formula 2.12).
- At an arbitrary point, the EMF has a shape of the TEM wave (formulas 2.13 and 2.14).
- Vectors **E** and **H** can have two spatial components that are shifted in phase. As a result the field is elliptically polarized.
- The dependence of E and H from φ and q is described by the normalized directional pattern that is independent of R.
- The E and H components are mutually perpendicular and proportional while the proportionality factor is equal to the wave impedance of a medium.
- The Poynting vector **S** = **E** × **H** is real and oriented radially.

To characterize the EMF properties in a far-field, we have presented a simple example of Maxwell's equations. To get a generalized solution of the equations, it would be necessary to take into account the diffraction of a wave caused by irregularities in a non-homogeneous medium, dispersion and non-linear properties of the medium, its anisotropy as well as the superposition of waves when a non-monochromatic field is being considered. The general solution of the Maxwell equations is still unknown. We shall see that, both from the point of view of these considerations and in the majority of cases crucial for metrological practice, a general solution is not necessary.

2.2. The far-field and the near-field

The considerations presented above lead us to the description of several features that characterize the far-field. There is no limit, in the sense of a discontinuity, between the far-field, the intermediate-field, and the near-field. However, in order to distinguish the specificity of the near-field and to create an approximate delimitation of the far-field and the near-field, one of the criteria for their delimitation is presented below [2].

If we calculate the difference between the distance r given by (2.6) and its approximate magnitude given by the first two terms of the series in (2.7), then multiply the difference by k, we will have a relationship describing the phase error $\Delta\Psi$ in elements of integration in formulas (2.3) and (2.4). The limits of the use of the approximation R>>D are defined by the error and may be expressed in the form:

$$\Delta\Psi = k\left(\frac{R'^2}{2R}\sin^2\beta + \frac{R'^3}{2R^2}\cos\beta\sin^2\beta + \cdots\right) \quad (2.16)$$

If we accept the maximal magnitude of R' and assume 2R' = D, then we obtain the maximal value of the error:

$$\Delta\Psi_{max} = \frac{kD^2}{8R} = \frac{2\pi}{N} \quad (2.17)$$

where N = a number depending upon an acceptable inaccuracy of the phase front. Usually it is assumed that $N \geq 16$, then:

$$R \geq \frac{2D^2}{\lambda} \quad (2.18)$$

This condition is widely accepted as a limit of the far-field. To illustrate it, let's consider two examples relating to antennas working at different frequencies and having different sizes:

- The limit of the far-field of an antenna with a parabolic reflector of 3 meters in diameter working within a 10 GHz band,
- The limit of the far-field of the tallest antenna in the world, a long-wave transmitting center in Gabin, Poland with a height a bit above 0.5λ, operating at 227 kHz.

In both cases, the far-field begins at distance above about 600 meters away from the antenna. If in our consideration three terms of series (formula 2.7) are taken into account, i.e.:

$$r = R - R' \cos \beta + \frac{R'^2}{2R} \sin^2 \beta \qquad (2.19)$$

and then similar considerations are repeated. We obtain the following condition:

$$R \leq \frac{D}{4} + \frac{D}{2}\left(\frac{D}{\lambda}\right)^{1/3} \qquad (2.20)$$

where Formula 2.20 gives the limit of the near-field.

Figure 2.2 shows (after [2]) modes of the field around an aperture antenna. In Figure 2.3, the locations of the near- and far-field boundaries are indicated as a function of r, l, D and λ. The near-field and the intermediate-field are referred to as the Fresnel region (Fresnel zone), while the far-field is referred to as the Fraunhofer region or the radiation field. When in close proximity to a radiation source, where the field may be assumed as the stationary one and E is independent of H conversely, the behavior is defined as an induction field. (Instead of using Maxwell's equations here, the use of the Biot-Savart law and Coulomb's law are assumed to be sufficient.) Here the imaginary part of the Poynting vector is dominant.

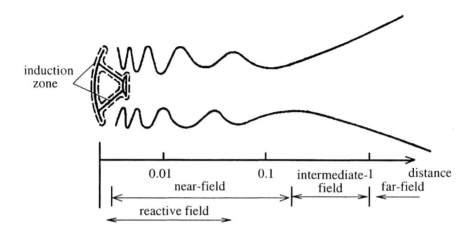

Figure 2.2. *EMF in the proximity of an aperture antenna.*

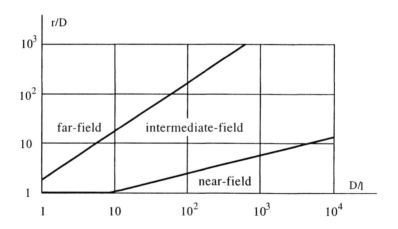

Figure 2.3. *EMF near a source as function of r, l, D and λ.*

The formulas introduced to define the near-field boundary (2.20) and the far-field boundary (2.18), require a word of comment. The series expansion given by Formula 2.7 is true if R' < R, or more precisely, if:

$$R' < R(\cos\beta + \sqrt{\cos^2\beta + 1})$$

Although the conditions are not always fulfilled, formulas 2.18 and 2.20 are widely applied in the literature as definitions of the far-field and the near-field limits. The accepted approximation is a result of arbitrarily assumed permissible nonhomogeneity of the phase front N. On the borders there appears no discontinuity of the EMF characterizing vectors and the expression "border" was introduced here in order to systematize the EMF parameters in the region surrounding a source. We may add here that the above definitions of the boundaries are not the only ones. The criterion may be based on, for instance, the convergence of the E/H ratio to Z_0, the Poynting vector to the electric (magnetic) power density and others, but they are more difficult for making a precise determination for a general case as compared to the ones presented. Nevertheless, any criterion is based upon arbitrarily chosen values of a parameter and the choice may be difficult to justify (e.g. why we accepted N = 16 instead of 15 or 17).

While spatial EMF components in the near-field are calculated, the rigorous use of the general dependencies (for instance, formulas 2.3 and 2.4) is indispensable and appropriate precautions should be taken when any simplifications in calculations are planned. A special caution is necessary when applying software for numerical analysis without appropriate analysis of the simplifications and assumptions that have been accepted in the procedures.

As noted earlier in Section 2.1, properties of EMF in the far-field appear partly in the intermediate-field as well, although none of them appear in the near-field. This results in the necessity of the specific methods used for EMF measurements in both regions. Several examples are quoted below to illustrate this point:

- In the far-field, E and H measurements are fully equivalent and they permit the calculation of the other component as well as S. In the near-field, separate E and H component measurements are indispensable, and they remarkably complicate the issue of the S measurement,
- The EMF polarization in the near-field, especially in conditions of multipath propagation, may be quasi-ellipsoidal because of the spatial orientation variations of the polarization ellipse. This is due to, for instance, the frequency of a source variation as a result of its FM modulation, Doppler effect due to reflection from a moving object, etc.,
- The radiation pattern in the far-field is constant and independent of the distance to a source; near-field measurements on the ground may be calculated only for sources of regular structure using complex computations [4],
- The Poynting vector in the near-field is complex and its direction and magnitude are functions of the source structure and the distance to the source.

2.3. EMF from simple radiating structures

If in formulas (2.3) and (2.4) we assume that the electric current has a non-zero magnitude in the direction of axis z, i.e. *$\mathbf{J}$ = 0 and $|\mathbf{J}| = J_z$ = const, for: $-l/2 < z < +l/2$, and at the same time $l << \lambda$, R (where: l is the length of the dipole's arm), then using the calculated Π, we substitute into formulas (2.1) and (2.2) solve for the components of the EMF generated by an electric elemental dipole placed in the Cartesian coordinate system as shown in Figure 2.4.

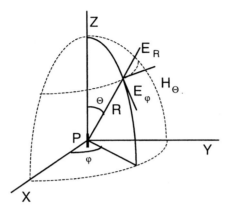

Figure 2.4. *Elemental electric dipole in Cartesian coordinates.*

The components are:

$$E_R = \frac{p}{2\pi\varepsilon}\left(\frac{1}{R^3} - \frac{jk}{R^2}\right)\cos\theta\, e^{-jkR} \tag{2.21}$$

$$E_\theta = \frac{p}{4\pi\varepsilon}\left(\frac{1}{R^3} - \frac{jk}{R^2} - \frac{k^2}{R}\right)\sin\theta\, e^{-jkR} \tag{2.22}$$

$$H_\varphi = \frac{j\omega p}{4\pi}\left(\frac{1}{R^2} - \frac{jk}{R}\right)\sin\theta\, e^{-jkR} \tag{2.23}$$

where p = the dipole moment:

$$p = |\mathbf{p}| = \left|\frac{I_z\, dz}{j}\mathbf{1}_z\right| \tag{2.24}$$

and I_z = the current in the dipole
  $\mathbf{1}_z$ = the versor of axis z.

If we repeat similar procedures for $J = 0$ and $|{}^*\mathbf{J}| = {}^*J_z =$ constant, we will have formulas defining the spatial field components of the elemental magnetic dipole. Using the principle of duality, the formulas may be immediately obtained from (2.21), (2.22) and (2.23) by the exchange of E to H and ε to μ and conversely.

The maximal spatial variations of the EMF components, as a function of distance to a point of observation, may be written in the following forms:

For an elemental electric dipole:

$$E = C_1(\alpha) \frac{\exp(-jkR)}{R^\alpha} \qquad (2.25)$$

$$\text{while } 3 \geq \alpha \geq 0, \text{ for : } 0 < R < \infty \qquad (2.26)$$

where $C_1(\alpha)$ is a constant dependent of R, and $E = |E|$ (complex amplitude of the electric field strength).

And by analogy, for an elemental magnetic dipole:

$$H = C_2(\alpha) \frac{\exp(-jkR)}{R^\alpha} \qquad (2.27)$$

where $C_2(\alpha)$ is a constant dependent of R, and $H = |H|$ is the complex amplitude of the magnetic field strength, while α and R fulfill conditions (2.26).

Formulas (2.25) and (2.27) illustrate the curvature of the EMF and are true with no regard to the zone. The radius of curvature is proportional to $R^{\alpha+2}$ and varies from infinity (for the plane wave) to very small magnitudes in the near-field.

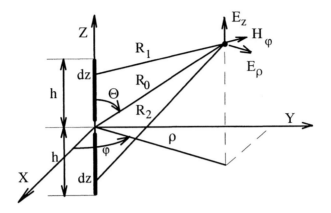

Figure 2.5. *Symmetrical dipole in coordinate system.*

If we substitute $|\mathbf{J}| = J_z \sin k(h - |z|)$, and $^*\mathbf{J} = 0$ in formulas (2.3) and (2.4) within the limits $-h \leq z \leq +h$ (where h is the length of the dipole arm), after calculation of Π using the formulas (2.1) and (2.2), we find components of the EMF from an infinitely thin, symmetrical dipole antenna of total length 2h (Figure 2.5). As a result of the sinusoidal current distribution assumption in the dipole a certain error is accepted. The error is especially important for not ideally thick (one dimensional) dipoles; however, the assumption is fully acceptable while the radiation pattern of such antenna is being considered. The use of precise solutions of the integral equations, describing current distribution in a real antenna, is not necessary in the aspect of the EMF components' strength calculations in the area of interest as well as in light of the final conclusion of the considerations presented [5]. We will use here the results of calculations available in the literature [1]. The EMF components are given by the following formulas (2.28), (2.29) and (2.30) respectively:

26 ELECTROMAGNETIC FIELD MEASUREMENTS IN THE NEAR FIELD

$$E_z = -\frac{jZI_0}{4\pi \sin kh} \left[\frac{\exp(-jkR_1)}{R_1} + \frac{\exp(-jkR_2)}{R_2} - 2\cos kh \frac{\exp(-jkR_0)}{R_0} \right]$$

$$E_\rho = \frac{jZI_0}{4\pi\rho \sin kh} \left[\frac{\exp(-jkR_1)}{R_1}(z-h) + \frac{\exp(-jkR_2)}{R_2}(z+h) - 2\cos kh \frac{\exp(-jkR_0)}{R_0} z \right]$$

$$H_\varphi = \frac{jI_0}{4\pi\rho \sin kh} \left[\exp(-jkR_1) + \exp(-jkR_2) - 2\cos kh \exp(-jkR_0) \right]$$

where: I_0 = current intensity at the dipole input,

$$R_1 = \sqrt{\rho^2 + (z-h)^2}$$
$$R_2 = \sqrt{\rho^2 + (z+h)^2}$$
$$R_0 = \sqrt{\rho^2 + z^2}$$

other indications as in Figure 2.5.

If we continue with considerations similar to the above, in the case of the electric and magnetic elemental dipoles, and generalizing, on the basis of formulas (2.28), (2.29) and (2.30), we can write per analogy to formulas (2.25) and (2.27) the relations describing the EMF variations in proximity to the symmetrical dipole:

$$E = C_3(\alpha) \frac{\exp(-jkR)}{R^\alpha} \qquad (2.31)$$

$$H = C_4(\alpha) \frac{\exp(-jkR)}{R^\alpha} \qquad (2.32)$$

where $1 \geq \alpha \geq 0$ and

$C_3(\alpha)$ and $C_4(\alpha)$ = constants dependent of R
R = the distance to a point of observation, while EMF close to the antenna surface is considered one may assume: $R \approx \rho$.

If, in formulas (2.25) and (2.27) we substitute $\alpha = 3$, we will have a relationship defining the EMF variations as a function of distance in the near-field of elemental dipoles where $\alpha = 2$ represents the intermediate-field of the dipoles. The far-field of the elemental dipoles and the near-field of a thin symmetrical dipole antenna are characterized by $\alpha = 1$. The variability of the latter versus distance is specific to a spherical wave. Rigorous analysis of formulas (2.21) to (2.23) and (2.28) to (2.30) does not justify an assumption in the formulas that $\alpha = 0$ for $R \Rightarrow \infty$, which would represent the plane wave. Such a simplification is often accepted when an EMF in a limited area, sufficiently far from a source, is being considered. In that area, amplitude variations of E and H vectors in any direction are negligibly small. The simplification is equivalent to the assumption that the radius of curvature of the field considered is equal to infinity. The maximal phase variations are independent of α if one assumes that α is a constant, such a case is most interesting from the point of view of metrological practice.

The comparison of formulas (2.25) and (2.27) as well as (2.31) and (2.32) permits us to come to the conclusion, which is very important for near-field EMF metrology, that the EMF "variability" in the proximity of sources much smaller in the comparison to the wavelength ($\alpha = 3$) is the largest. Thus, if we estimate the errors of the EMF measurements near the sources, the majority of the errors will be for an arbitrary source. The conclusion is, in some sense, an intuitive one and it is a result of the presence of the quasi-stationary field in proximity to sources whose sizes are comparable or larger than the wavelength (induction field). One example of this is EMF that surrounds AC power devices and especially overhead transmission lines. While the fields are being analyzed, the validity of Maxwell's equations is often "forgotten" and only Coulomb's Law and Biot-Savart's law are applied. The approach is equivalent to the assumption that the EMF does not exist and the field is sufficiently represented by E and H fields only.

Doubt may arise under these considerations, relating to the presence of higher powers when a multipole expansion is applied. The approach makes it possible to obtain more precise calculations of EMF generated by elemental sources. However, even if appropriate corrections are applied, it does not make remarkable changes in relation to the majority of errors. It is especially true when arbitrarily small physical sources are considered. It should be emphasized that only physical sources have a practical importance because of the efficiency of the EM energy radiation. Good examples are formulas describing the standard EMF near a standard loop antenna. In this case, apart from the finite sizes of the antenna that are remarkably larger than the elemental dipole ones, there does not exist a term in power exceeding 3. Again this is a matter of practical importance [6].

2.4. Bibliography

1. D. J. Bem, *Antennas and Radiowave Propagation* (in Polish). Warsaw 1975.
2. B. Minin, *VHF Radiation and the Human Security* (in Russian). Moscow, Sovetskoe Radio 1974.
3. Ju. D. Dumanskij, A. M. Serbyuk, I. P. Los, *The Influence of RF Electromagnetic Fields on Humans* (in Russian). Kiev 1975.
4. P. F. Wacker, *Non-planar Near-Field Measurements: Spherical Scanning,* National Bureau of Standards, Publ. NBSIR 75-809, Boulder, CO USA.
5. A. Karwowski, P. Buda, *The Method of the Protection Zones in Proximity of Medium- and Long Wave Transmitting Antennas* (in Polish). Prace IL No. 93/87, pp. 2–27.
6. H. Trzaska, *Magnetic Field Standard at Frequencies Above 30 MHz,* HEW Publications, (FDA) 77-8010, vol. II, pp.68–82, Rockville MD.

3

EMF Measurement Methods

In order to select an optimal method for EMF measurement in the near-field, it is first necessary to determine which quantities best characterize the field. These quanitites will then be the subject of the measurement.

From the point of view of antenna performance evaluation, it is essential to measure the strength of E or H components near the antenna, which then makes it possible to find the current or the charge distribution along the antenna. With this as a basis, it is possible to find the radiation pattern of the antenna and its input impedance. The measurement of E, H or S in the near-field (with the phase information conserved) permits, with some complex calculations, finding the antenna's radiation pattern in the far-field. From the point of view of shielding, absorbing, or EMF attenuating materials, investigations of the E, H and S measurements are sufficient.

If we are interested in protection against unwanted exposure to EMF, and in biomedical investigations in particular, the E, H and S measurements are not enough. This area of investigation requires more precise qualification of the parameters that should be a subject of the measurement. The proposals cited previously for the protection standards provide, as the basic criterion of the interaction of the EMF with biological media, the power or energy absorbed in the mass unit (Specific Absorption Rate (SAR) and Specific Absorption (SA) in Table 1). Sometimes the power absorbed in volume units is applied, and a widely accepted measurement is the temperature rise due to EM energy absorption (which permits the determination of SAR and SA). The measurement of current induced in a body by external EMF has recently become more popular. The majority of protection standards require the measurement of the root-mean-square (RMS) value that reflects the quantity of the absorbed energy. In the nonthermal approach, it is more important to know the amplitudes of the field components, their spatial positioning and their temporal variations, as well as the frequency of the carrier wave and that of the modulating spectra (and their temporal variations) and the type of modulation. Although this approach is presently unfashionable, in the author's opinion the nonthermal data will be the primary future requirement.

3.1. E, H and S measurement

In Chapter 1, portions of several versions of protection standards were presented to illustrate the range of measured magnitudes of E, H and S. Let's stress again that these magnitudes only show requirements for the surveying and monitoring services. Only laboratory experiments will require field measurements from the lowest measurable magnitudes (near the noise level or even below the noise level) to the highest which can be generated by the use of available power sources. Moreover, the levels given by the standards vary in the succeeding versions, modifications and actualizations of those standards. The other parameters of the measured field are much less rigorously defined in the standards. Let's consider them.

3.1.1. Spectrum of the measured EMF

At an arbitrary moment of time, in a chosen point in space there exists a solitary vector E and a solitary vector H. They are linearly polarized and their magnitude is equal to the sum of instantaneous values of any spatial components and spectral fringes appearing at the point considered. The conditions may be written in the form:

$$E = E_0 + \sum_{i=1}^{N} E_i \cos(\omega_i t + \varphi_i) \qquad (3.1)$$

where

E_0 = the electrostatic field strength
E_i = the strength of the i-th spectral fringe
ω_i = the angular frequency of the i-th fringe and
φ_i = the phase of i-th fringe.

If we substitute H instead of E in formula (3.1) we obtain the formula defining the temporal variations of the magnetic field. If we neglect the static component in the formula we note that, without regard to the region considered (Fresnel or Fraunhofer Region), and with the exception of guided waves, E is orthogonal to H. We should note here that the spatial positioning of the resultant vector is not given by the formula and the positioning may be arbitrary. The sum given by the formula (3.1) is a finite one. In many practical cases, N does not exceed one or two. However, even in the simplest cases, simultaneous measurement of all the frequency fringes may be technically difficult or even impossible, as in the case of simultaneous measurement of static and RF components. At times it may be undesirable because of interpretational problems. For instance, when the measurements conditions are such that the fringes fall in frequency ranges where different levels are permitted.

The issue has three important aspects:

1. It is technically possible to construct an EMF meter with a frequency response equivalent to the frequency limits of the protection standards. While the frequency response of a meter is a continuous function of frequency, the protection standards are characterized by discontinuities at the borders between

frequency ranges where the limits are different. The meters are compatible with one standard only and, sometimes not even in accordance with one standard (for instance where a work safety limit differs in its frequency shape with that of the general public). The meters, apart from their convenience for inspection officers, do not permit evaluation of the actual roles of the separate EMF sources, especially when they fall within different frequency ranges.

2. Wideband EMF measurement by inspection services is the most convenient technique because of the speed and simplicity of the measurement. In order to assure unequivocal results of the measurement, the use of a meter covering more than one frequency range, as represented in the standards, requires switching off any other source apart from the measured one. However, even in such a situation, the presence of spectral harmonics radiated by the source may lead to problems with interpretation of the results of such a measurement.

3. A selective measurement can also be troublesome, especially when measurements are performed in the presence of a large number of sources. However, the measurement allows precise estimation of the role of any separate EMF source in the resultant field. A new concept for such a measurement is presented in Chapter 9. The author began his involvement in the field with the selective meters designed in the early 1960s. After more than 30 years, in his opinion, the selective methods are most useful and may be considered as the best methods for the future.

3.1.2. EMF polarization

The expression *polarization* is understood in three ways:

1. As positioning of the vector E in relation to a chosen reference system, e.g., vertical and horizontal polarization in relation to the Earth's surface,

2. As the shape of an envelope of the E (or H) vector rotations in the space (linear polarization, circular polarization, elliptical or quasi-ellipsoidal),

3. As the direction of the E (or H) vector rotations in the space (left- and right-hand polarization).

For our consideration, taking into account the first two meanings of polarization is enough.

The maximal value of E and H does not result from polarization in any above sense, whereas the RMS value depends on the polarization only in the sense defined in point 2 above — where the magnitude of the energy absorbed by a body, for instance, as well as the current induced by an EMF in the body, are a function of the field vectors' positions in relation to the body. In the latter case, the polarization sense of points 1 and 2 is of concern. This shows the importance of the EMF polarization field and the necessity of its measurement.

It is necessary to call attention here to the dependence of the results of EMF measurements on the polarization of the measured field and the directional pattern of the probe applied. We must understand the advantages and disadvantages of probes with sinusoidal, circular and spherical directional pattern when an EMF of an unknown polarization is being measured. The problem will be briefly discussed in Chapter 7.

3.1.3. EMF Modulation

Each quantity represented in formula (3.1) may be the subject of intentional variation as a function of time (modulation). The variation of E is called its amplitude modulation (AM), and a very important type of amplitude modulation is pulse modulation. When the subject of variation is ω we call it frequency modulation (FM), and φ alternation is called phase modulation (PM). The carrier wave modulating signal may be analogue or discrete.

When monochromatic fields are being measured there are few problems with the results of the measurement interpretation. In the case of a modulated field measurement, and in particular when pulsed fields are measured, the question arises: what do we measure? Do we measure the maximal instantaneous value, mean value or RMS? The answer to the question is still somewhat in doubt and should be given by biologists and medical doctors based on detailed studies of the importance of thermal interactions (RMS measurement) or non-thermal ones (peak value measurement). The role of an engineer should be an auxiliary one as a consulting support during laboratory studies and as the person responsible for making appropriate choices for measuring devices or an exposure system fulfilling the requirements of the experiment.

Let's to focus our attention on a technical aspect of the problem. In further considerations it will be shown that the RMS value is measured by a probe using a square-law detector. The indication of the meter is proportional to the RMS value of the sum of any spectral fringe in a particular frequency band. However, the design of a correctly functioning square-law detector is difficult, especially when the probe is intended to work in the near-field, in wide frequency range, and with large dynamic range. Such a probe has not yet been constructed. The measurement of the instantaneous peak value, especially of short monopulses, is extremely troublesome and its realization requires the use of expensive measuring devices and complex analytical methods to reconstruct the shape of the measured pulse. It is possible that the standards should suggest (or even require) the simultaneous measurement of both values, however, it will increase the cost of the measurements as well as make them more burdensome. In order to simplify the measuring procedures, as well as to decrease costs of the typically expensive measuring equipment, the majority of meters available on the market provide for measurement of the EMF of an uninterrupted envelope, although the limitation is rarely mentioned in the manuals of these devices.

The measurements of E, H and S are usually achieved with the use of probes based upon an electric or magnetic antenna of small electrical sizes and loaded with a diode detector. Detailed considerations related to these probes are presented in Chapter 4 and the following, while magnetostatic fields or very large field (VLF) magnetic fields are usually measured with the use of Hall-effect devices or other types of semiconductor devices. Although their detailed analysis is not taken into account, some of the considerations presented here may be helpful when these sensors are applied.

3.1.4. The use of the far-field meters for the near-field measurements

The basic features of near-field EMF measurement devices are the small size (both in the physical and electrical sense) of a measuring probe as well as potentially poor directional properties. Less evident, is the necessity of using electric field sensors based on electric antennas (whip, symmetric dipole antenna) and sensors with magnetic antennas (loop, ferrite rod) for the magnetic field measurement. An exception to this rule will be discussed in Chapter 6, as confusion may arise from the widespread use of the meters equipped with loop antennas (because of their better stability, reduced sensitivity to the presence of conducting objects in their proximity and relatively smaller sizes) and calibrated in E-field units. The latter may be used for the far-field measurements *only*, where the constant relation between electric and magnetic field is valid, as given by formulas (2.13) or (2.14). The phenomenon (as evident) will not be discussed in the farther parts of the work. It requires, however, a few words of comment as even people experienced in EMF measurements often make such a mistake (gross error). The power density S is also often measured by the way of E or H measurement. While this is very true in the far-field, it requires knowledge and caution when such a measurement is performed in the near-field. It should not be necessary to add that power density meters, equipped with resonant-size antennas (horn or log-periodic antennas) in the near-field measurements are completely useless. Remember the far-field boundary estimations presented in the previous chapter: the far-field limit of a relatively small sized parabolic antenna may exceed several hundred meters, or more!

In the near-field, the mutual relationship between both field components is unknown *a priori*. The relationship depends upon the structure of the source of radiation, and it is a function of direction and distance between the point of observation and the source. The exception to the rule, expressed by formulas (2.13) and (2.14) and valid only in the far-field, is sometimes used as one possible criteria for the far-field boundary [1].

Criteria for evaluation of the measuring antenna's size and the directional pattern of the probes used are subjects of detailed analysis presented in later chapters. They are mentioned here only in order to focus our attention on the most important features of the meters used in near-field metrology.

3.2. Temperature rise measurements

The temperature rise measurement makes it possible to evaluate SAR:

$$\text{SAR} = \frac{C_p \Delta T}{t} \left[\frac{W}{kg} \right] \quad (3.2)$$

where

ΔT = measured temperature increase [K],
C_p = specific heat of the absorber (phantom) [kJ/kgK],
t = exposure time [s].

In order to illustrate the energy transfer from the EMF to a phantom, consider the simplest case of a lossy dielectric inserted between plates of a capacitor. Without taking into account the heat transfer to the surroundings (by radiation or conduction), i.e., taking into account only the thermal capacity of the body, in the conditions of full thermal insulation, the time to warm the body is:

$$t = \frac{\rho\, C_p\, \Delta T}{\varepsilon_o\, \varepsilon_r\, \omega\, E^2\, \text{tg}\, \delta} \qquad (3.3)$$

where ρ = mass density [kg/m^3].

If in formula (3.3) we substitute the mean magnitudes of the living tissues' parameters and we assume the minimal measurable increase of temperature $\Delta T \approx 0.1$ K and E = 10 V/m, then for such idealized conditions (without taking into consideration the heat transfer!) the time required for the temperature rise (in 0.1 K) is 10^5 to 10^{10} seconds, depending upon frequency of the field. Sensitivity is the most important factor limiting the method application in EMF measurements.

The temperature measurement may be characterized by the following comments:

- The temperature rise represents the largest quantity of EMF energy that can be absorbed by a body, a good agreement between theoretical analyses and practical experimentation is obtained with no regard to the field modulation, polarization, etc.
- Because of thermoregulation mechanisms *in vivo*, the relation between the measurements *in vitro* and phenomena *in vivo* creates some variations. Similar variations arise due to the heat transfer from the body to its surroundings,
- The exposure measurement using a phantom permits preservation of the full analogy between the conditions of measurement in relation to an exposed person (the analogy does not exist in any other measuring method — the large size of the 'probe,' normally one of the its most important inconveniences, may be considered here as an advantage),

- The use of such a 'meter,' especially for our purposes, seems to be extremely troublesome,
- More difficult, as compared to other methods, is obtaining results that are repeatable, univocal and comparable with the measurements carried out in different conditions and with the use of other methods,
- The measurements performed *in vivo*, although possible, are burdened with a substantial measurement error resulting from the impossibility of differentiating bioeffects caused by EMF exposure and the necessary damage caused by the placement of a sensor in a living body,
- There is no realistic possibility of distinguishing between the polarization of the EMF illuminating the absorber (phantom) or its frequency. Because of the thermal inertia, it is not possible to measure the EMF modulation,
- The frequency response of the phantom is a function of its dimensions and shape, as well as a function of the position of the phantom in relation to EMF vectors,
- As mentioned above, the sensitivity of the method is not sufficient,
- There is also doubt if and how to measure the average value of the absorbed energy for the whole body (phantom), versus a point value.

The methods of temperature measurements are well known from technical publications. However, because of the perspective attractiveness in the field, especially in some laboratory applications, as well as possible wide spread use of the technique in the future, for microwave EMF measurements in particular, selected methods of the temperature measurement, worked out with regard to the EMF measurement needs are presented below.

3.2.1. Temperature measurement with the use of liquid crystals

The essence of the method is based upon the investigation of the tincture or the light reflection coefficient of a liquid crystal. The crystal is immersed in a micro container and illuminated with the use of an optical fiber, another fiber leads light reflected from the surface of the crystal to a photodetector.

It is possible to achieve here the resolution of 0.1 K with frequent calibration of the sensor. The resolution decreases to about 0.25 K without the calibration because of thermal drift and aging of the crystal.

Because of the absence of a conducting component in the device's design, the sensor (including feeders) is 'transparent' to the measured field, which eliminates measured field disturbances and, as a result, increases the accuracy [2].

3.2.2. Temperature measurement with the use of a thermoelement

The method is based upon the use of a thermoelement immersed in a thin-walled glass pipe, which is then inserted into the tissue under investigation. In order to limit disturbances of the measured field by metal leads of the thermoelement as well as eliminate the possibility of EMF penetration into the tissue, the measurement is performed before and after exposure of the tissue. While it is being exposed the thermoelement is withdrawn from the pipe [3].

3.2.3. A thermistor temperature measurement

The use of a thermistor inserted into tissue allows continuous observation of temperature variations while the tissue is exposed to the EMF. The errors of measurement, due to the field deformations caused by the thermistor and by its metallic leads, can be mitigated by using a high resistance thermistor and resistive leads which are transparent to the measured field. The resistance variations are measured by a bridge.

The method was modified to increase sensitivity and accuracy of the measurement [4]. For the purpose of the experiment, a high resistance thermistor and transparent leads (160 kΩ/cm) were used. The latter carry a 0.3 µA DC exciting current. An additional pair of leads, connected directly to the thermistor, permits measurement of the voltage drop across the thermistor. The DC power dissipated by the thermistor does not exceed 0.1 µW while the DC power in the leads is less than 0.05 µW/cm. The resolution of the device exceeds 0.01 K.

3.2.4. Temperature measurement with the use of a viscosimeter

Variations in liquid viscosity as a function of temperature allow the use of this phenomenon for temperature measurements. The presented device [5] includes a system of capillary tubes throughout which a liquid is pumped. The pressure difference in a capillary at the input of the sensor and at its output is a measure of the temperature. The pressure difference is measured by a transducer. The measuring ranges and the deivce sensitivity change are achieved by the liquid choice.

3.2.5. Thermographic and thermovisional measurements

The development of theoretical analyses of absorption models has lead to the 'millimeter resolution' models [6]. Although the models are not the subject of the work we are reminded that the first model studies, initiated by Guy and Johnson, were followed by experimental studies which made it possible to verify the theory and the correctness of the necessary simplifying assumptions in it. The measurements were performed applying thermovision and thermograph cameras. The experimental models, of different geometrical shapes and sizes, contained several parts that permitted observation of different model cross sections and as a result, the temperature distribution in the sections after the model exposure. The models were usually electrically homogeneous and isotropic. However, they allowed measurement of many interesting results showing the dependence among shape, size, and electric properties of the model (phantom) on one hand and the manner of exposure, frequency, EMF polarization in relation to the object and modulation on the other [7]. A good agreement between the theoretical and the experimental results has in many cases lead to the renunciation of the latter one as more expensive and troublesome when compared to the model studies.

These methods have allowed quick, simple and easy visualization of the temperature distribution in a chosen plane of the model and, as a result, localization of the thermal extrema (hot spots) while different combinations of exposure are being used. A disadvantage of the method is its thermal inertia and the necessity of needing visibility of the investigated area. On the other hand, its doubtless advantage is the possibility distance measurement, with no physical contact between a sensor and a body, (remote sensing) which permits limited disturbances of EMF in proximity to the exposed body as well as useful measurement results for their archivization and computer analysis.

3.3. Current measurements

Contrary to the temperature measurements, the measurement of the current induced in a human body by the EMF, is already the subject of acting legal regulations. The measurement is especially useful as a measure of the hazards created by EMF at the lowest frequency ranges, in particular in the neighbourhood of the overhead high-voltage transmission lines, and power substations as well as near long- and medium-wave broadcast high power transmitters, where polarization parameters of the measured EMF are well known. An additional advantage of the measurement (and especially corresponding to it legal regulations) is the possibility of including EM radiation hazard and the electric shock in one protection standard.

The current measurement is taken one of three ways. By placing a person on a conducting, standard size plate and measuring the current between the plate and the surface (of the earth) using a thermocouple; measuring a voltage drop on a resistance between the plate and the ground, or by using a current transformer (Figure 3.1).

42 ELECTROMAGNETIC FIELD MEASUREMENTS IN THE NEAR FIELD

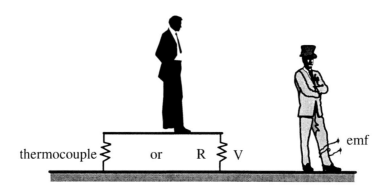

Figure 3.1. *Methods of measurement of the current flowing through a human body.*

Contrary to the electric shock, the use of the current measurement for an EM radiation hazard raises the following uncertainties:

- The result of the measurement is dependent not only on the posture of the person being measured, but also the clothes they are wearing and in particular their shoes.
- The current measurement accuracy is extremely dissatisfying,
- The current measurement, in the manner shown in Figure 3.1, in accordance with standards in power, entirely reflects currents induced in the body by EMF components parallel to the vertical axis of the standing person. It is essential to note the impossibility of measuring the horizontal components of the current and not talk about currents induced in the body by H-field (eddy currents). These have been the subject of intense biomedical investigations lately.

Presented methods of measurement are concerned only with current measurements in the foot or the leg. It was shown, however, that remarkable currents may flow throughout other parts of the body or its organs. For example, a hand or lip current of someone using a walkie-talkie [8].

The results of hand current (I_h) measurements and the lip current measurements (I_l), for several types of hand-held radiotelephones with 5 W nominal output power and supplied with different antenna types, are shown in Table 3-1. The column DE shows increase of the E-field strength, measured at distance 10 meters from the device, while the radiotelephone is held in a hand by its operator in relation to the same device placed on an insulating support. The results shown in Table 3-1 emphasises the role of an operator's body as a "counterpoise," especially at lower frequencies.

Frequency	ΔE	I_h [mA]	I_l [mA]	antenna	power [W]
27 MHz	15x	170	120	1.5 m	5
27 MHz	10x	150	100	25 cm	5
144 MHz	3x	90	70	5/8λ	5
144 MHz	2x	80	70	15 cm	5
432 MHz	1.5x	50	50	12 cm	5

Table 8. *Hand and lip current of a radiotelephone operator.*

The role may be confirmed by the presence of the standing waves on the arm of an radiotelephone operator and the E-field maximum at distance of about λ/4 from the radiotelephone antenna's base as shown in Fig. 3.2. The experiment was repeated with walkie-talkies working within 150 and 450 MHz bands.

Figure 3.2. *A standing wave on the arm of a radiotelephone operator.*

Let's call attention to the fact that the maximal current intensity at the antenna base depends mainly on the device output power, as the antenna input impedance is usually a standardized one. The current may be estimated as follows:

$$I_{max} = \sqrt{\frac{P_{fed}}{R_{in}}} \qquad (3.1)$$

where

P_{fed} = power fed to the antenna
R_{in} = input impedance.

After we substitute into Formula 3.1 typical values for the devices used in the experiment, i.e.: $P_{fed} = 5$ W and $R_{in} = 50$ ohms we will have the maximal magnitude of the current intensity flowing through the hand of an operator:

$$I_{max} @ 0{,}3 \text{ A}$$

The estimated current intensity will appear at relatively low frequencies, where electrical sizes of the radiotelephone casing are much less than the wavelength. The measured currents are well below the 0.3 A, which may be the result of the role played by the casing, the measurement conditions or very limited accuracy of the measurement.

As shown in Table 3.1, the measured currents sometimes exceed the permissable magnitudes given in Tables 1.5 and 1.7 for a foot or leg! It is not this book's place to discuss the correctness (or incorrectness) of the accepted legal regulations. However, we have shown the necessity of accounting not only for the theoretically estimated EMF energy absorption from a radiotelephone antenna (as it was done till now), but also the conducted currents I_h and I_l while the energy absorption in an operator body is studied and the hazard created by these devices is investigated. Although biomedical interaction is not the subject of the work, it is worth mentioning two problems related to mobile communications:

- It is not clear if the use of a mobile radiotelephone (cellular phone) should be considered as a professional exposure or non-professional one
- The majority of the theoretical studies of EM energy absorption in a body of hand-held radiotelephones is devoted to calculation of the absorbed energy and its distribution in the body (head) while the possibility of the EMF envelope detection suggests the necessity of accounting for the role in vivo of detected VLF currents within the exposed body.

3.4. Bibliography

1. D. A. Tschernomordik , "Estimation of the Far-Field Boundary of a Symmetrical Dipole" (in Russian), *Trudy NIIR*, No. 4/1972, pp. 55–60.
2. C. C. Johnson, T. C. Rozzell, "Liquid Crystal Fiber Optic RF Probes, Part I," *Microwave Journal*, 1975, No. 8, pp. 55–57.

3. C. C. Johnson, A. W. Guy, "Nonionizing Electromagnetic Wave Effects in Biological Materials and Systems," *Proc. IEEE*, 1970, Vol. 60, pp. 692–718.
4. E. L. Larsen, R. A. Moore, J. Acevado, "A Microwave Decoupled Brain – Temperature Transducer," *IEEE Trans.*, Vol. MTT-22, 1974, pp. 438–444.
5. C. A. Cain, M. M. Chen, K. L. Lam, J. Mullin, *The Viscometric Thermometer*, US Dept. of Health, Education and Welfare HEW Publication (FDA) 78-8055, pp. 295–305.
6. J. Y. Chen, O. P. Gandhi, D. Wu, "Electric Field and Current Density Distributions Induced in a Millimeter-Resolution Human Model for EMFs of Power Lines," *XVIth Annual Meeting of the BEMS*, Copenhagen, 1994.
7. A. W. Guy, "Analyses of Electromagnetic Fields Induced in Biological Tissues by Thermographic Studies of Equivalent Phantom Models," *IEEE Trans.*, Vol. MTT-19, 1971, pp. 205–214.
8. H. Trzaska, "ARS and EM-Radiation Hazard," *Proc. 1994 Int'l. EMC Symp.*, Sendai, pp. 191–194.

4

Electric Field Measurement

The basic method of electric field measurement, within a wide frequency range, involves the use of a charges' induction in a body illuminated by the field. As shown in Figure 4.1, the electromotive force (emf), e_E, induced by the electric component of the EMF generated by an arbitrary source in a symmetrical dipole antenna of total length 2h is:

$$e_E = \int_{-h}^{h} \mathbf{E}\, \mathbf{dh} \tag{4.1}$$

If the source of the field is the elemental electric dipole placed in the coordinate system shown in Figure 2.4 and the length of dipole l fulfils the condition $l << R_0$, then making use of formula (2.25) we write:

$$e_E = C_1(\alpha) \int_{-h}^{h} \frac{\exp(-jkR)}{R^\alpha} \cos(\mathbf{E}, \mathbf{dh})\, dh \tag{4.2}$$

48 ELECTROMAGNETIC FIELD MEASUREMENTS IN THE NEAR FIELD

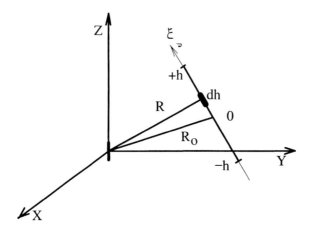

Figure 4.1. *A symmetrical dipole in a coordinate system.*

(4.2)

where R = the distance between the point of integration and the center of the coordinate system:

$$R = \sqrt{R_0^2 + \xi^2 + 2R_0\,\xi\,\cos(\mathbf{E},\mathbf{dh})} \qquad (4.3)$$

where

ξ = current length of the antenna, $-h \leq \xi \leq h$,
$\cos(\mathbf{E},\mathbf{dh}) = \mathbf{1_E} \times \mathbf{1_h}$
$\mathbf{1_E}$ and $\mathbf{1_h}$ = versors of vector $\mathbf{E}$ and $\mathbf{h}$ respectively.

Formula (4.2) allows an estimation of the magnitude of emf induced in the measuring antenna while the latter is arbitrarily located in the space relative to the field source. Under further consideration, the formula will be applied for E-field measurement error estimations, specifically for the near-field measurements and for the antenna (E-field probe) directional pattern synthesis.

4.1. Field averaging by a measuring antenna

The maximal changes of amplitude and phase of the measured field along an antenna will appear for the radial component of the field, i.e., for **E** parallel to **h**, then:

$$e_E = C_1(\alpha) \int_{R_0 - h}^{R_0 + h} \frac{\exp(-jkR)}{R^\alpha} dR \qquad (4.4)$$

Making use of the Schwartz inequality, after integration, we have (4.5):

$$e_E \leq C_1(\alpha) \sqrt{\frac{\exp(-j2kR_0) \sin 2kh}{k} \left[\frac{1}{1 - 2\alpha} \left(\frac{1}{(R_0 + h)^{2\alpha - 1}} - \frac{1}{(R_0 - h)^{2\alpha - 1}} \right) \right]}$$

If the measuring antenna fulfils the condition $h \ll R$ and $h \ll \lambda$, this represents the calibration's conditions, and the emf induced in the antenna e' is:

$$e'_E = C_1(\alpha) \frac{\exp(-jkR_0)}{R_0} 2h \qquad (4.6)$$

Dividing formula (4.6) by formula (4.5) gives:

$$\frac{e'_E}{e_E} \leq \sqrt{\frac{2kh}{\sin 2kh} \left[\frac{(1 - 2\alpha) 2h}{R_0^2} \frac{(R_0^2 - h^2)^{2\alpha - 1}}{(R_0 - h)^{2\alpha - 1} - (R_0 + h)^{2\alpha - 1}} \right]} \qquad (4.7)$$

or:

$$\frac{e'_E}{e_E} \leq \frac{kh}{\sin 2kh} + \frac{(1 - 2\alpha) h}{R_0^{2\alpha}} \frac{(R_0^2 - h^2)^{2\alpha - 1}}{(R_0 - h)^{2\alpha - 1} - (R_0 + h)^{2\alpha - 1}} \qquad (4.8)$$

50 ELECTROMAGNETIC FIELD MEASUREMENTS IN THE NEAR FIELD

The error of the E-field measurement resulting from averaging of the measured field by the measuring antenna we define as:

$$\delta_E = \frac{e_E - e'_E}{e_E} \leq \delta_{1E} + \delta'_{2E} \tag{4.9}$$

where δ_{1E} = an error expressing the field's phase changes along the antenna:

$$\delta_{1E} = \frac{1}{2}\left(1 - \frac{2kh}{\sin 2kh}\right) \tag{4.10}$$

and δ'_{2E} = an error representing the amplitude changes along the antenna:

$$\delta'_{2E} = \frac{1}{2}\left[1 - \frac{(1-2\alpha)2h}{R_0^{2\alpha}} \frac{(R_0^2 - h^2)}{(R_0 - h)^{2\alpha-1} - (R_0 + h)^{2\alpha-1}}\right] \tag{4.11}$$

Formula (4.9) reflects measurement errors caused by two factors limiting the accuracy, i.e., averaging the phase of the measured field along the antenna and averaging the amplitude. We will analyze briefly the first of them here while the latter will be discussed in more detail in section 4.3 of this chapter.

The possibility of the field spatial distribution measurement, especially under conditions of multi-path propagation and interference, and the measurement of the maximal and minimal magnitudes of the field strength (while standing waves appear) requires the use of probes equipped with antennas whose sizes are much less than the wavelength. While it would be most convenient to use 'point antennas' (zero-dimensional ones), but since they are unavailable, we will consider the results of formal, though slightly overestimated calculations of the measurement errors due to finite electrical length of the E-field probe's antenna. Formula (4.10) gives the error resulting from the phase changes along the antenna and dependent from kh. The formula makes it possible to find the magnitude of kh for which the value of the error will not exceed a permissible level. The error δ_{1E} is plotted in Figure 4.2 as a function of kh.

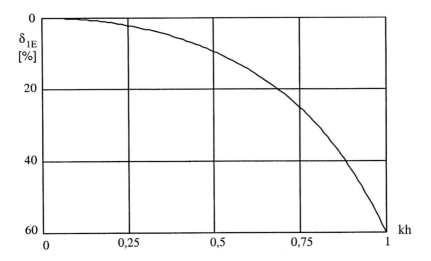

Figure 4.2. *Error δ_{1E} versus kh.*

It should be noted here that the estimated magnitude of the error shown in Figure 4.2 is about twice that defined by the ratio of the current intensity at the input of the antenna to that averaged along the antenna. In light of the above discussion, the use of antennas longer than h < λ/4 seems to be unacceptable. In the radio frequency ranges and lower, the antenna length limit presented may be seen as evident exaggeration, as the measuring antenna, even one much shorter than a quarter-wave, can exceed the size of the space where the measurement is to be performed. We may conclude that the electrical sizes of the antennas will become important when we are concerned with measuring decimeter-waves and shorter.

4.2. Influence of fields from beyond a probe measuring band

As mentioned in Chapter 3, for ease of measurement, it is most advantageous to use wideband meters. However, because of both the accuracy of the monochromatic field measurement and the measurement of the field whose spectrum is unknown, it is important to know, if possible, the precise frequency response of the probe.

The typical wideband probe used for electric field measurement consists of a symmetrical dipole antenna loaded by a detection network. For instance, a diode, a thermocouple or an optical modulator, whose output voltage leads to an indicating device. As has been shown [1], with no regard to the type of detector applied, the probe may have certain maxima in its frequency response, especially at the highest frequencies. They result from the increase of the effective length of the antenna, resonance of the probe's parasitic reactance, alternations of the detector efficiency with frequency and others. Meanwhile, the frequency response of the probe should be maximally flat within the entire measuring band (medium frequency range) and it is assumed that outside the band the sensitivity of the probe should not exceed that within the band. The latter suggests the use of RC low-pass filters that would artificially cut off the response at the highest frequencies on one side and permit its shaping, if desired, on the other. Figure 4.3 is shows a simplified schematic diagram of the probe described. The probe is additionally equipped with the low-pass filter mentioned, which is placed between the antenna and the detection diode and allows the probe's response shaping. Figure 4.4 presents its equivalent network for high frequencies.

ELECTRIC FIELD MEASUREMENT 53

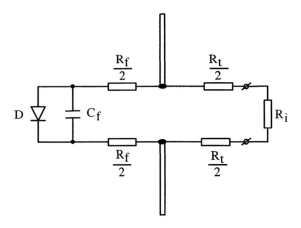

Figure 4.3. *Simplified schematic diagram of the E-field probe.*

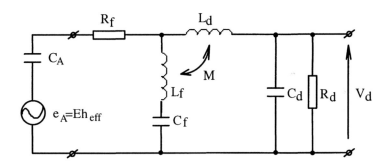

Figure 4.4. *Equivalent circuit of the E-field probe for high frequencies.*

If one assumes that, for dipole antennas with arms having a length of (h < λ/6), the input capacitance of the antenna (C_A) and its effective length (h_{eff}) are frequency independent, then the transmittance (T(jf)) of the probe, defined as the ratio of the voltage delivered to the detector (V_d) to the emf (e_E) induced by the field in the antenna, is given by:

54 ELECTROMAGNETIC FIELD MEASUREMENTS IN THE NEAR FIELD

$$T(jf) = \frac{V_d}{e_E} = \frac{1 - \left(\frac{f}{f_3}\right)^2}{\frac{C_f}{C_A} + 1 - \left(\frac{f}{f_1}\right)^2 - \left(\frac{f}{f_2}\right)^2 + \left(\frac{R_f}{R_d} + \frac{C_d}{C_A}\right)\left[1 - \left(\frac{f}{f_4}\right)^2\right] + \left(\frac{f}{f_1}\right)^2\left(\frac{f}{f_2}\right)^2}$$

$$+ j\left\{\left(\frac{f}{f_6} - \frac{f_5}{f}\right)\left[1 - \left(\frac{f}{f_4}\right)^2\right] + \frac{f}{f_7} + Q_d\left[1 - \left(1 - K^2\right)\left(\frac{f}{f_2}\right)^2\right]\right\}^2 \quad (4.12)$$

where

$$f_i = \frac{\omega_i}{2\pi} \quad i = 1-7$$

$$\omega_1^2 = \frac{1}{L_d C_d}$$

$$\omega_2^2 = \frac{1}{L_f C_f}$$

$$\omega_3^2 = \frac{1}{C_f (L_f + M)}$$

$$\omega_4^2 = \frac{1}{C_f (L_f + L_d + 2M)}$$

$$\omega_5 = \frac{1}{C_A R_d}$$

$$\omega_6 = \frac{1}{C_d R_f}$$

$$\omega_7 = \frac{1}{C_f R_f}$$

$$Q_d = \frac{\omega L_d}{R_d}$$

$$K = \frac{M}{\sqrt{L_f L_d}}$$

R_d = parallel connection of the diode resistance ρ and $R_t + R_i$, and the other indications are the same as in Figure 4.3, in particular:

R_f, C_f = components of the RC low-pass filter
R_t = resistance of the transparent line
R_i = input resistance of a DC voltmeter
C_d = equivalent capacitance of the detector

Formula (4.12) allows the analysis of the probe transmittance in different configurations. The formula may be reduced if we assume $M = 0$, which is usually valid. The parasitic inductances may be omitted and this does not affect the transmittance, especially when it is limited at the highest frequencies. The resistance of the transparent line as well as that of the voltmeter may sometimes be neglected; they physically do not exist, for instance, when an optical modulator loads the antenna. It makes it possible to apply the formula when it is in its simplified form for the transmittance estimations of different versions of the E-field probe.

The transmittance given by (4.12) covers a wide frequency range that may be divided into three frequency bands:

1. The low frequency band, where:

$$|T(jf)| = \frac{f}{f_s} \qquad (4.13)$$

2. The medium frequency band (measuring band) with constant transmittance:

$$T(jf) = \text{const} = \frac{C_A}{C_A + C_d + C_f} \qquad (4.14)$$

while corner frequencies of the band, at which the transmittance decreases by 3 dB, are defined as follows:

The lower corner frequency: f_l, for $f \to 0$, while R_d is equal to the equivalent reactance of the probe:

$$f_l = \frac{1}{2\pi R_d (C_A + C_d + C_f)} \qquad (4.15)$$

or if we assume the transmittance's decrease in any degree δ:

$$f_l(\delta) = \frac{1-\delta}{2\pi R_d (C_A + C_d + C_f)\sqrt{1-(1-\delta)^2}} \qquad (4.15a)$$

The upper corner frequency: f_u, for $f \to \infty$:

$$f_u = \frac{C_A + C_d + C_f}{2\pi R_f C_A (C_d + C_f)} \qquad (4.16)$$

or, per analogy to (4.15a), for the transmittance reduction in δ:

$$f_u(\delta) = \frac{\sqrt{(1-\delta)^2 - 1}\,(C_A + C_d + C_f)}{2\pi(1-\delta)\,R_f C_A (C_d + C_f)} \qquad (4.16a)$$

Formula (4.16) is true if $R_f \ll R_d$ and $f \ll f_{1-4}$,

3. The high frequency band, where $f \gg f_{1-4}$:

$$|T(jf)| = \frac{1}{f}\frac{f_6 f_7}{f_6 + f_7} \qquad (4.17)$$

Formula (4.17) reflects the issue in far-simplified form because it does not take into consideration the changes of the antenna's electric parameters, within which the frequency band may take place. However, it reveals the possibility of achieving the decreasing run of the transmittance modulus in a chosen frequency range if a low-pass filter (a detector of shaped frequency response) is used.

Even if we take into consideration the foregone simplifications, applied when formulas (4.15) and (4.16) were introduced, they are of primary practical importance. The formulas make it possible to select the detector and filter parameters that allow us to obtain the flat characterizations of the probe's frequency response within the desired frequency band using a single low-pass band filter. In particular, they enable the selection of corner frequencies and the transmittance limitation in the upper band. The use of multiple filters enables shaping of the probe's frequency response within the thresholds of a protection standard. However, it must be stated that the use of any filter reduces the probe's sensitivity.

Now we will outline more general considerations. As shown above, our measurements can be applied only to antenna sizes that are much smaller in comparison to the shortest wavelength of the measuring band. From both the possibility of precisely calculating the probe's frequency response in the whole frequency range and, in particular, at frequencies above the measurement band, and the possibility of using the probe at frequencies exceeding those accepted for its use in the far-field, we will also analyze the transmittance of the probe for $h > \lambda/6$. Additionally, we will consider the advantages that may result from using an n-segment RC filter. The transmittance of such a probe, whose equivalent network is shown in Figure 4.5, is given by formula (4.18).

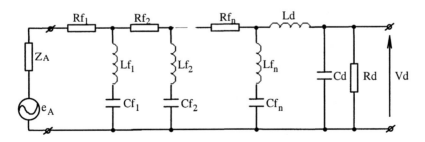

Figure 4.5. *Equivalent network of the probe with an n-segment RC filter.*

$$T = \frac{V_d}{E} = \frac{h_{eff}}{A_{11}} \qquad (4.18)$$

where A_{11} = term of a matrix A given by:

$$[A] = [A_A] \cdot [A_{f1}] \cdot [A_{f2}] \cdots [A_{fn}] \cdot [A_d] \qquad (4.19)$$

and

$$[A_{fi}] = \begin{bmatrix} 1 + \dfrac{j\omega R_{fi} C_{fi}}{1 - \omega^2 L_{fi} C_{fi}} & R_{fi} \\ \dfrac{j\omega C_{fi}}{1 - \omega^2 L_{fi} C_{fi}} & 1 \end{bmatrix} \qquad (4.20)$$

Index 'i' denotes i-th segment of the filter, $1 \le i \le n$

$$[A_d] = \begin{bmatrix} 1 + \dfrac{j\omega L_d}{R_d}(1 + j\omega R_d C_d) & j\omega L_d \\ \dfrac{1 + j\omega R_d C_d}{R_d} & 1 \end{bmatrix} \qquad (4.21)$$

$$[A_A] = \begin{bmatrix} 1 & Z_{11} \\ 0 & 1 \end{bmatrix} \qquad (4.22)$$

where Z_{11} = input impedance of an antenna given by (4.33).

The effective length of a thin symmetrical dipole antenna with a large length to diameter ratio is [2]:

$$h_{eff} = \frac{2[J_o(kh) - \cos kh]}{k \sin kh} \qquad (4.23)$$

where J_o = Bessel function of the first kind and zeroth order.

Examples of calculated frequency responses of the E-field probe are shown in Figure 4.6. The curves are normalized in relation to that for n = 3.

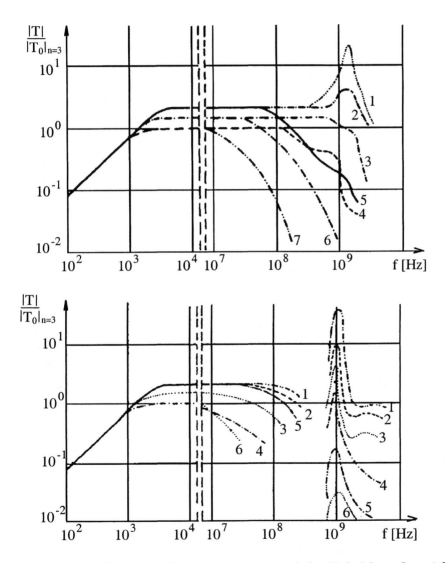

Figure 4.6. *Calculated transmittances of the E-field probe with n-segment filters.*

The upper diagram shows results of calculations performed with an assumption of $L_f = 0$ whereas the lower diagram is for $L_f \neq 0$. Curve 1 shows the transmittance of the probe without a filter, curves 2 and 5 represent the probe with a single-segment filter, curves 3 and 6 refer to a double filter, and curves 4 and 7, a triple one. The calculations were done for similar time constants of the filters.

From the curves shown in Figure 4.6, we can draw the conclusion that the use of multi-segment filters makes it possible to reach the desirable frequency response shape and large attenuation within the high frequency range. The conclusion is evident in some sense, but its use is strongly limited. Primarily because of a decrease of the probe's sensitivity in proportion to the number of filter segments in the applied probe. The transmittance of the probe with an n-segment filter, within the medium frequency band, is given by:

$$T(jf) = \frac{C_A}{C_A + C_d + \sum_{i=0}^{n} C_{fi}} \tag{4.24}$$

Thus, the larger the sum of the filter's capacitances, the smaller the transmittance and, consequently, the meter's sensitivity. Even a single-segment filter in the probe permits us to obtain its transmittance shape at the highest frequencies such that the sensitivity, above an arbitrarily selected upper corner frequency, won't exceed that within the measuring band. Simultaneously, we must remind ourselves that using even a single-segment filter causes a reduction in the probe's sensitivity that results directly from formulas (4.14) and (4.24).

The results of the above calculations represent one case of the cascade connection of several identical filters. The use of multi-segment filters of different resonant frequencies permits us to design the frequency run of the transmittance similar to that of the frequency dependent thresholds given by a protection standard. The filters also allow construction of a probe with several bands in which the transmittance will be frequency independent and of different (desirable) magnitudes.

By using the mentioned filters as well as traps, tuned to the antenna's resonant frequencies, it is possible to design a super wideband E-field probe that would have a flat frequency response at frequencies corresponding to the length of the applied antenna limited by h < 0.5λ. Although it would be possible to successfully continue this approach to obtain an acceptably flat response above the top limit, the directional pattern of the antenna splits and the number and magnitude of the lobes in the pattern increases thereby excluding the use of such antenna (with no regard, of course, to the previous discussion of the antenna's size limitation). Even if the shape of the directional pattern is in some cases permissible, the synthesis of an E-field probe with a spherical directional pattern (omnidirectional) and acceptable pattern irregularities, seems impossible.

A schematic diagram and its equivalent circuit of a probe with a trap are shown in Figure 4.7, whereas the calculated frequency responses are shown in Figure 4.8. Curve 1 shows the transmittance while the trap was not tuned to the antenna's resonant frequency and the quality factor of the trap was too high. In Curve 2, the trap was tuned but still with unchanged quality factor. Curve 3 represents optimal compensation of the resonant effect. The probe may be used principally when superwideband measurements are performed in the far-field and with the use of a panoramic receiver or a spectrum analyzer, when the detection diode is replaced by a light modulator.

62 ELECTROMAGNETIC FIELD MEASUREMENTS IN THE NEAR FIELD

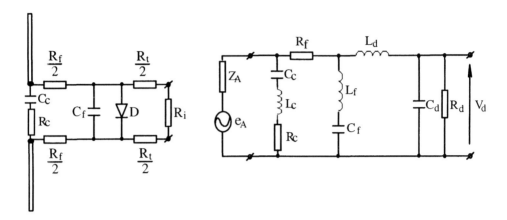

Figure 4.7. *Schematic diagram and equivalent network of the E-field probe with an RC filter and a trap $R_c C_c$.*

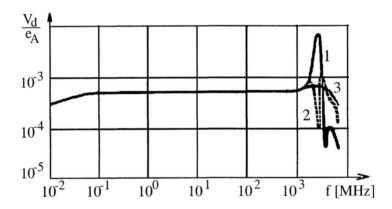

Figure 4.8. *Normalized transmittance of the probe with an RC filter and a trap.*

In the above examples, the detector was represented by linear elements. It made it possible to simplify the analyses, but at the expense of precise measurements. This reservation reflects mainly the diode detector, a bit less than with the thermocouple and it does not concern the optical modulator. Currently, however, the diode detectors are the most popular choices in this field. Because of the dependence of the nonlinear element parameters on its chosen working point, especially the lower corner frequency it may be a function, among others, of the intensity of the measured field [3]. However, the main aim here was to show the possibilities and necessities of the probe's frequency response shaping and prospective errors that could result when performing measurements with probes of uncontrolled run response, even if during the measurements, only a single source is input. The errors may be caused by the harmonic frequencies' presence in the spectrum radiated by the source or the power-line frequency field and radiation caused by other unexpected sources (for instance: ultrasonic generators, HF power sources, excited high power lasers, information equipment, nearby BC and TV stations) or that were not switched off. They continue generating fields that could interfere with the measured one and consequently, may affect measurement results.

Here an additional comment is indispensable. Some meter manufacturers suggest in their manuals, the possibility of using the meters at frequencies far below their lower corner frequency and preparing the measurement results by multiplying the meter readout by a manufacturer-provided correction factor taken from the frequency response. Almost all these meters are equipped with probes that include nonlinear detectors (diode or thermocouple one). Usually, while no filters are in the probe or a single filter is used in the low frequency range, the slope of the transmittance is 6 dB/octave. The slope is valid for a linear detector (as it was assumed in the above calculations of the frequency response) whereas for a square-law detector the slope will be twice as much. While the detector works in its linear or non-linear range, it is impossible to determine whether or not an error occurs. Therefore, it is imperative to drop this approach. The gross error may be the result of the completely impossible evaluation of the importance of the source's harmonics in the meter's indication (the sensitivity, in the range, increases proportionally to the harmonic's order or to its square); the mentioned dependence of the frequency response from

64 ELECTROMAGNETIC FIELD MEASUREMENTS IN THE NEAR FIELD

the amplitude of the measured field and the detector's working point; the temperature dependence of the detector's parameters and others. These problems will be discussed in greater detail in Chapter 8.

4.3. Mutual interaction of the measuring antenna and the field source

4.3.1. The dependence of the accuracy of E-field measurement from the structure of the source

The measurement accuracy limitations due to field averaging by the measuring antenna were discussed in 4.1. The equations defined there allowed us to the define the errors δ_{1E} and δ'_{2E}. The former was a function of the h/λ ratio and represented the influence of the phase variations along an antenna while the latter was dependent on R_0/h and α. Thus, it represents, in the accuracy of measurement estimations, both the relative distance between the antenna and the source; and the structure of the source. As mentioned earlier, because of simplifications accepted when formula (4.11) was introduced, the error calculated using the formula was almost overestimated by a factor of two. In order to make these estimations more precise, we will repeat the calculations assuming (with an accuracy to δ_{1E}) that the phase variations of the measured field along the antenna do not take place, in other words, the term $\exp(-jkR)$ in Formulas (4.4) and (4.5) is equal to $\exp(-jkR_0)$. It means that the measuring antenna is short enough (in the wavelength measure) and that the emf induced in the antenna is averaged along the antenna only as a result of a finite curvature of the measured field.

$$e_E = C_1(\alpha)\exp(-jkR_0)\int_{R_0-h}^{R_0+h}\frac{dR}{R^\alpha} \qquad (4.25)$$

and

$$e'_E = C_1(\alpha) \exp(-jkR_0) \frac{2h}{R_0^\alpha} \tag{4.26}$$

Making use of formulas (4.25) and (4.26), we now define the error δ_{2E} as:

$$\delta_{2E} = \frac{e_E - e'_E}{e_E} \tag{4.27}$$

Substituting formulas (4.25) and (4.26) in to (4.27) we have for $\alpha = 3$:

$$\delta_{2E} = 1 - \frac{\left(R_0^2 - h^2\right)^2}{R_0^4} \tag{4.28}$$

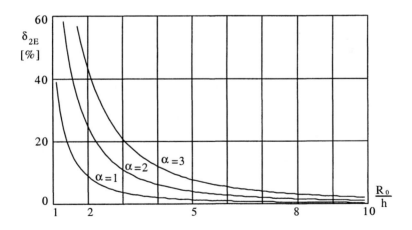

Figure 4.9. *The error δ_{2E} as a function of R_0/h for $\alpha = 2$.*

$$\delta_{2E} = \frac{h^2}{R_0^2} \qquad (4.29)$$

and for $\alpha = 1$

$$\delta_{2E} = 1 - \left[\frac{R_0}{2h} \ln \frac{R_0 + h}{R_0 - h}\right]^{-2} \qquad (4.30)$$

With no regard as to how we approach the calculations for the plane wave, for which $\alpha = 0$, we have $\delta_{2E} \equiv 0$.

The run of the error, calculated using formulas (4.28), (4.29) and (4.30), is shown in Figure 4.9. The curves illustrate the influence of the structure of the source upon the accuracy of the E-field measurement as a function of R_0/h. We must stress here that the effect of averaging the measured field limits the size of the antenna designated for the measurements in close proximity to a primary or secondary source the most rigorously. The resultant accuracy of the field measurement in the near-field, in the majority of cases, is dominated by the error. The error is characterized by its rapid decrease with increasing distance as it may be deduced directly from Figure 4.9.

4.3.2. The dependence of the accuracy from the antenna input impedance changes

The mutual impedance of a thin conducting dipole of diameter 2a, located parallel at distance b/2 to a flat, infinitely large and perfect conducting plane, and its mirror reflection is given by Formula (4.31) [4]:

$$\begin{aligned}Z_{21} = 30 \Big[&\left(2 + e^{-j2kh}\right)(\text{Ci } kx - j \text{ Si } kx)\Big|_{\sqrt{h^2 + b^2} - h}^{b} + \\ &+ \left(2 + e^{j2kh}\right)(\text{Ci } kx - j \text{ Si } kx)\Big|_{\sqrt{h^2 + b^2} + h}^{b} + \\ &+ e^{j2kh}(\text{Ci } kx - j \text{ Si } kx)\Big|_{\sqrt{h^2 + b^2} + h}^{\sqrt{4h^2 + b^2} + 2h} + \\ &+ e^{-j2kh}(\text{Ci } kx - j \text{ Si } kx)\Big|_{\sqrt{h^2 + b^2} - h}^{\sqrt{4h^2 + b^2} - 2h} \Big] \qquad (4.31)\end{aligned}$$

Taking into consideration the limit:

$$\lim_{b \to a} Z_{21} = Z_{11} \qquad (4.32)$$

and taking into account that the sizes of the problem are much less when compared to the wavelength of the considered field, i.e., for a, b, h << λ, neglecting the real part of the impedance, as much smaller when compared to the imaginary one, we have the input reactance of a short dipole in the free space X_{11}. After taking the limit of the impedance given by formula (4.31) for λ→∞ and omitting its real part, we receive the mutual reactance of two dipoles placed at distance b apart. The demanded input reactance X_i of the dipole placed parallel to the perfectly conducting medium we receive as a difference of the two reactances, that is:

$$X_i = X_{11} - X_{21} \qquad (4.33)$$

For the free space, i.e., a space without any material objects that could affect the input capacitance of the antenna, $X_{21} \equiv 0$, the transmittance of the E-field probe, within the measuring band, given by formula (4.14) now we rewrite in the form:

$$T(jf) = \frac{X_d}{X_d + X_{11}} \qquad (4.34)$$

where X_d denotes the equivalent reactance represented by the detection network, i.e., the input reactance of the detector, the sum of reactances of the filters applied, and the parasitic capacitances of the probe.

If the influence of the material media upon the antenna's input reactance can not be neglected, in other words when $X_{21} \neq 0$, the transmittance is:

$$T'(jf) = \frac{X_d}{X_d + X_i} \qquad (4.35)$$

68 ELECTROMAGNETIC FIELD MEASUREMENTS IN THE NEAR FIELD

Because the probe is calibrated under the conditions where $X_{21} \approx 0$, when the measurements are performed in the neighborhood of a conducting medium (represented, in the estimations carried out, by the infinitely large and perfectly conducting plane) there will be a measurement error, caused by the probe's antenna input impedance change which is affected by the presence of the medium. We will define the error in the form:

$$\delta_{3E} = \frac{T'(jf) - T(jf)}{T'(jf)} \qquad (4.36)$$

The input impedance of the antenna as well as the mutual impedance of the antenna and its mirror reflection is a function of its slenderness ratio [2]. The slenderness depends on the ratio of the antenna length to the diameter of the conductor from which it's made, i.e., h/a. Thus, it is necessary to introduce the slenderness as a parameter in our calculations. Figure 4.10 shows calculated runs of the error δ_{3E} versus b/2h for h/a = 30, 300 and 1000 respectively.

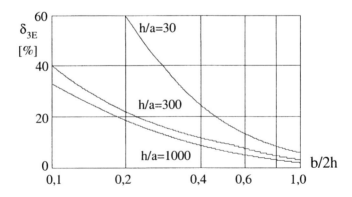

Figure 4.10. *Error δ_{3E} versus b/2h.*

Let's focus our attention on three features that characterize the results of our considerations:

- The errors δ_{2E} and δ_{3E} are functions of the distance probe-source, but their comparison shows that the error δ_{3E} is usually exceeded by δ_{2E}
- The above comment becomes more essential if we take into account that the estimations were carried out for an antenna located parallel to an infinitely large conducting medium. However, when near the conducting medium (represented in the calculations as a primary or a secondary source of radiation) the dominant role is played by the radial E-field component and the measuring antenna should be spatially oriented in accordance to the component. Thus, the presented estimations give results maximizing any error of this kind that could appear in the metrological practice
- The latter is valid even if we take into account that estimations of δ_{2E} were completed for an infinitely thin dipole and the magnitude of the error will increase for dipoles of finite thickness

These considerations may be summarized as the estimated magnitudes of errors in relation to those that may appear during the measurements are exaggerated. Thus, the estimation should be understood as an illustration of possible maximal errors representing the inaccuracy of measurement caused by separate limiting factors.

4.4. Changes to the probe's directional pattern

The precision of EMF measurements in the near-field, near numerous primary and secondary sources, where there are usually three spatial field components, i.e., quasi-spherical polarization, includes the necessity of simultaneous measurement of all three components. We should explain at the very beginning that spherical field polarization, in light of physics, does not exist; and that the quasi-spherical or the quasi-spherical polarization is understood here as a circular or an elliptical polarization whose plane of polarization makes arbitrary rotations in space. This may occur for

70 ELECTROMAGNETIC FIELD MEASUREMENTS IN THE NEAR FIELD

instance, as a result of multi-path propagation and interference. These measurements may be achieved using a meter equipped with a probe and single dipole antenna, whose directional pattern is of sinusoidal shape, by way of individual spatial component strength measurements and then by finding the total field using simple calculations. However, because of the difficulties in making large numbers of calculations, this procedure may lead to gross errors and mistakes.

By intuition it may be surmised that design of a probe which is sensitive to the three spatial components of the field should include an antenna system containing three mutually perpendicular dipole antennas as shown in Figure 4.11.

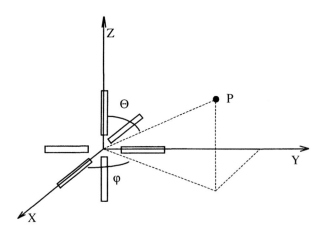

Figure 4.11. *A set of three mutually perpendicular dipoles.*

The emf induced in separate dipoles by an arbitrarily located E-field vector may be calculated using Formula (4.1). For i-th dipole it is:

$$e_{Ei} = \int_{-hi}^{hi} \mathbf{E} \cdot \mathbf{dh}_i \tag{4.37}$$

If, in formula (4.37) we substitute 'i' for x, y and z we solve for the electromotive forces induced each of the three dipoles. If we assume then, that the antennas are of equal size, i.e., $h_i = h_x = h_y = h_z$, then they are much shorter than the wavelength and they are illuminated by a homogeneous monochromatic plane wave, with an accuracy to a constant A. The sum of the voltages induced in these antennas may be expressed in the form:

$$e_E = \sum_{i=x}^{z} e_{Ei} = A\left[\cos(\mathbf{E}, \mathbf{h}x) + \cos(\mathbf{E}, \mathbf{h}y) + \cos(\mathbf{E}, \mathbf{h}z)\right] \quad (4.38)$$

Because the formula should be valid for any spatial localization of the vector **E** let's assume, for instance, that its direction coincides with that of the polar vector **R** of the spherical coordinate system (Figure 4.1). Thus, the directional cosines of the vector **E** are identical with those of vector **R**, i.e., cos q, sin q cos φ and sin q sin φ.

When arbitrarily polarized field measurements are performed, we want the probe to be insensitive to the kind of polarization and the direction of the **E** (or **H**) vector spatial rotations. In order to fulfil this condition, we need the emf induced in the measuring antenna to be independent of the spatial coordinates of the **E** or **H** vector. For the antenna system shown in Figure 4.11 we may initially suppose that the presented requirements are to be fulfilled as separate antennas and that they should be sensitive to specific spatial components of the **E** vector. To show that this is true, it is enough to prove that the sum of the directional cosines in the square brackets of formula (4.38) is a constant value. We will use a simpler method: we will equate the formula (4.38), after its mentioned modification is completed, to zero. In a non-trivial case ($A \neq 0$), if the hypothesis is true, the following equation must not be fulfilled:

$$\cos\theta + \sin\theta\cos\varphi + \sin\theta\sin\varphi = 0$$

72 ELECTROMAGNETIC FIELD MEASUREMENTS IN THE NEAR FIELD

This formula presents an equation of a plane, which includes the center of the coordinate system and also intersects all the axes of the system at identical angles. Any change of the **E** vector localization (which is equivalent to the change of signs of separate terms in the equation), in relation to the coordinate system, results only in the spatial rotation of the plane. Thus, there is a non-vanishing set of uni-planar **E** vectors on the plane that are 'invisible' to the considered system of three mutually orthogonal antennas whose output voltages are directly added. It is obvious however, that the separate terms in the formula added after squaring result in unity. That means that the sum is independent of the spatial coordinates of the considered vector. This approach answers our search for the directional pattern of a probe that is insensitive to the spatial location of the **E** vector, i.e., an omnidirectional or spherical directional pattern.

Chapter 7 is devoted to a discussion of the synthesis of the spherical pattern probes. Here we will analyze the errors resulting from pattern changes that occur when an omnidirectional probe is used to perform measurements in the near-field.

Standardized directional patterns of three identical dipoles, whose arms are of length h, located on separate axes of a Cartesian coordinate system (Figure 4.11) are given by:

$$f_x(\theta, \varphi) = \frac{\cos(kh \sin\theta \cos\varphi) - \cos kh}{\sqrt{1 - \sin^2\theta \cos^2\varphi}} \quad (4.39)$$

$$f_y(\theta, \varphi) = \frac{\cos(kh \sin\theta \sin\varphi) - \cos kh}{\sqrt{1 - \sin^2\theta \sin^2\varphi}} \quad (4.40)$$

$$f_z(\theta, \varphi) = \frac{\cos(kh \cos\theta) - \cos kh}{\sin\theta} \quad (4.41)$$

If the parameters of the dipoles are identical, they are loaded by the square-law detectors of identical efficiency (that may be briefly said: three identical probes located on three axes of the coordinate system) then by summation of their output voltages we calculate the directional pattern of the system $F(q, \varphi)$:

ELECTRIC FIELD MEASUREMENT 73

$$F(\theta,\varphi) = f_x^2(\theta,\varphi) + f_y^2(\theta,\varphi) + f_z^2(\theta,\varphi) \qquad (4.42)$$

We define measurement errors resulting from irregularities of the system pattern, in the form:

$$\delta_{4E} = 1 - \frac{F_{min}}{F_{max}} \qquad (4.43)$$

where F_{min} and F_{max} are the minimal and the maximal value of the function given by formula (4.42) respectively.

The magnitude of the calculated directional pattern irregularities versus kh using formula (4.43), is shown in Figure 4.12. The considerations we've presented here are related to the electrical length (kh) of the dipoles and, thus, to the field phase variation along the dipoles, which is of special concern when resonant antennas are used (for shorter antennas, the error vanishes). In our field, such antennas may sometimes be used, especially at microwave frequencies and in some applications relating to electromagnetic interference metrology. Based on formula (4.43) and Figure 4.12, we can estimate the maximum permissible electrical length (h/λ) of the antennas; for shorter antennas the irregularities of the pattern should not exceed those assumed.

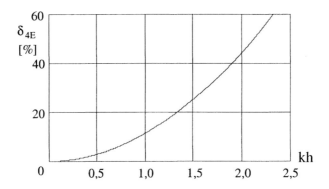

Figure 4.12. *Error d_{4E} versus kh.*

Considerations relating to omnidirectional pattern irregularities are similar to the above-presented estimations for the measured field averaged by a measuring antenna. As before, we may assume that for an antenna fulfilling the condition h << λ, the phase changes along the antenna (here with an accuracy to δ_{4E}) are of negligible importance with regard to the deformations of the directional pattern. Further considerations will be limited to the influence of the field amplitude changes along the antenna. Making use of Formula (4.2) we describe the emf induced by a point source located at point P in three mutually perpendicular dipoles whose directions coincide with those of separate axes of the Cartesian coordinate system. Thus:

$$e_{Ex} = C_1(\alpha) \exp(-jkh) \int_{-h}^{h} \frac{\cos(\mathbf{E, hx})}{R_x^\alpha} dx \quad (4.44)$$

$$e_{Ey} = C_1(\alpha) \exp(-jkh) \int_{-h}^{h} \frac{\cos(\mathbf{E, hy})}{R_y^\alpha} dy \quad (4.45)$$

and

$$e_{Ez} = C_1(\alpha) \exp(-jkh) \int_{-h}^{h} \frac{\cos(\mathbf{E, hz})}{R_z^\alpha} dz \quad (4.46)$$

where e_{Ex}, e_{Ey} and e_{Ez} = emf induced by the field in dipoles located on axes x, y and z that represent non-standardized directional patterns of the three short dipoles, and R_x, R_y, R_z = distance between point P (R_0, q, φ) and dx, dy and dz respectively:

$$R_x = \sqrt{R_0^2 + x^2 - 2R_0 x \sin\theta \cos\varphi}$$

$$R_y = \sqrt{R_0^2 + y^2 - 2R_0 y \sin\theta \sin\varphi}$$

$$R_z = \sqrt{R_0^2 + z^2 - 2R_0 z \cos\theta}$$

If we sum the squares of the emf induced in the three dipoles, we obtain the emf e(q, φ), which represents a non-standardized directional pattern of the system:

$$e(\theta, \varphi) = e_{Ex}^2 + e_{Ey}^2 + e_{Ez}^2 \qquad (4.47)$$

The synthesis of the spherical pattern of the system in the far-field (or under calibration conditions) is determined by the above-mentioned parameters, the identity of the three antennas, and loading of the detectors. We should assume that the conditions can be fulfilled with the required accuracy. The pattern deformation, resulting from the specific conditions of the near-field measurement, in particular from the finite curvature of the field, we define (with accuracy to δ_{4E}) as:

$$\delta_{5E} = 1 - \frac{e_{min}}{e_{max}} \qquad (4.48)$$

Calculated deformations of the pattern δ_{5E} as a function of R_0/h for α = 1, 2 and 3 are plotted in Figure 4.13.

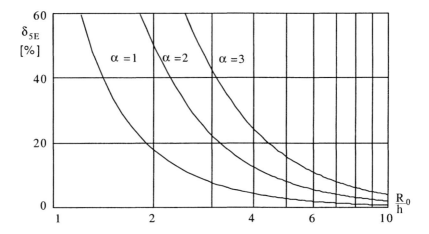

Figure 4.13. *Error δ_{5E} versus R_0/h.*

The pattern is also affected by the input impedance changes of the separate antennas that are a result of their proximity to a conducting medium. In order to estimate the effect, it would be necessary to follow the considerations presented earlier in section 4.3.2 for an omnidirectional probe. However, the section concluded with a comparison of the errors δ_{2E} and δ_{3E} and with the conclusion that δ_{2E} usually exceeds δ_{3E} by a considerable amount. Thus, we assume that in the case of the pattern deformations near media material, a similar conclusion is valid and we will omit further discussion of the factor. However, the reader should remember that it does not mean that the phenomenon does not exist.

Ending the considerations we notice the presence of the following relationships:

$$|\delta_{4E}| \approx 0{,}5\,|\delta_{1E}|$$

and

$$|\delta_{5E}| \approx 2\,|\delta_{2E}|$$

These relationships are self-evident as the pairs of errors are caused by the same physical phenomena. In other words, the phase averaging along a measuring antenna and the amplitude averaging by it, and caused by the effects on identical antennas, but in different spatial configurations.

4.5. The E-field probe comparison

A probe designated for wideband EMF measurements in the near-field, should fulfill the following requirements:

- The size of the probe's antenna should be considerably smaller in comparison to the minimal wavelength of the frequency range being measured, and they should not exceed (at first approximation) the minimal distance probe-source, at which the measurement is to be performed
- The transmittance of the probe should be as high and flat as possible within the measuring frequency band while the magnitude of the transmittance outside of the band should not exceed that within the band

and sometimes:

- The lower corner frequency should be as low as possible.

The first two first requirements contradict one another which creates the necessity of finding a compromise solution. For instance, an increase in antenna size results in a sensitivity increase; however, the size increases result in an increase of errors caused by the mutual interaction of the antenna and the field source. As a result, there is a decrease in the permissible upper corner frequency of the antenna-equipped probe and the distance probe-source where the measurement may be performed. If its desired accuracy has to be unchanged, the upper corner frequency must be enlarged.

Both the necessity of comparing probes of different parameters and then optimizing them requires the introduction of a common measure which takes into account the above mentioned three postulates. We call this measure the quality factor (q) of the probe and we define it in the form:

$$q = \frac{T(jf)}{f_1} \qquad (4.49)$$

As an example of the proposed approach to the field probes comparison we will present the possible application of the measurement technique for evaluating the applicability of three different types of antennas in the probe, i.e.: a thin symmetrical dipole antenna, the same with a capacitively loaded top (double T antenna) and a biconical antenna. Because the magnitudes represented in formula (4.49) are functions of the antenna length h (the condition: h << λ must be fulfilled) and the detector input impedance (R_d, C_d) we will assume in our estimations that they are identical for any probe considered. As a reference we will use the probe with the thin dipole antenna, whose quality factor is q_d, with gain g_c resulting from the extension of the dipole with its capacitive loading, of quality factor q_c, in relation to the probe with dipole, we define in the form:

$$g_c = \frac{q_c}{q_d} \qquad (4.50)$$

Analogously we define the gain, g_b, of the probe with a biconical antenna as:

$$g_b = \frac{q_b}{q_d} \qquad (4.50a)$$

Substituting Formulas (4.14) and (4.15) into Formula (4.49), we have:

$$q = 2\pi h_{eff} C_A R_d \qquad (4.51)$$

The quality factor of the probe with a dipole antenna is:

$$q_d = 2\pi h^2 R_d \frac{10^{-12}}{3{,}6 \, (\ln h/a - 1{,}7)} \qquad (4.52)$$

In the case of a dipole antenna with a total length 2h made of a conductor of diameter 2a and extended by a loading capacitance of the length h_1 and the conductor diameter $2a_1$, its electrical parameters are given by the following formulas:

1. The input capacitance C_i:

$$C_i = C_A + C_1 \qquad (4.53)$$

where C_A = input capacitance of dipole without the capacitive extension and C_1 = equivalent capacitance of the extension

2. Effective length:

$$h_{eff} = h \, \frac{h + 2h_1}{h + h_1} \qquad (4.54)$$

After substituting formulas (4.53) and (4.54) into formulas (4.49) and (4.50) we have the quality factor q_c and the gain g_c of the probe equipped with the capacitive extended symmetrical dipole:

$$q_c = q_d \, \frac{h + 2h_1}{h + h_1} \left(1 + \frac{h_1}{h} \, \frac{\ln h/a - 1{,}7}{\ln h_1/a_1 - 1{,}7} \right) \qquad (4.55)$$

and

$$g_c = \frac{h + 2h_1}{h + h_1} \left(1 + \frac{h_1}{h} \, \frac{\ln h/a - 1{,}7}{\ln h_1/a_1 - 1{,}7} \right) \qquad (4.56)$$

In the case of a biconical antenna of total length 2h and the cone's apex angle Θ_b, its quality factor q_b and the gain g_b we calculate using the following formulas which respectively gives the input capacitance C_i and the effective length h_{eff} of the short biconical antenna:

$$C_i = \frac{h\left(1 + \dfrac{\ln 4}{2 \ln \operatorname{ctg} \Theta_s/2}\right)}{3{,}6 \ln \operatorname{ctg} \Theta_b/2} \tag{4.57}$$

$$h_{eff} \approx h(1 + \sin \Theta_s/2) \tag{4.58}$$

The substitution of formulas (4.57) and (4.58) into formulas (4.49) and (4.50a) gives:

$$q_b = 2\pi h^2 R_d (1 + \sin \Theta_s/2) \frac{1 + \dfrac{\ln 4}{2 \ln \operatorname{ctg} \Theta_s/2}}{3{,}6 \ln \operatorname{ctg} \Theta_s/2} \tag{4.59}$$

and

$$g_b = \frac{(1 + \sin \Theta_s/2)\left(1 - \dfrac{\ln 4}{2 \ln \operatorname{ctg} \Theta_s/2}\right)\left(\ln \dfrac{h}{a} - 1{,}7\right)}{\ln \operatorname{ctg} \Theta_s/2} \tag{4.60}$$

In Table 9 are the set-up calculated and measured magnitudes of the quality factor and gain of the probes with the mentioned antennas. Both the calculations and the measurements were performed for $C_d = C_m + C_f$ (where C_m = equivalent capacitance of the detector that includes the probe's parasitic capacitances; for the investigated set it was estimated: $C_m \approx 1.8$ pF, while the capacitance of the detector itself was about 0.8 pF) and $C_f = 0$, 10 and 20 pF. The detector's equivalent resistance R_d was estimated as equal to 2 MΩ. The measurement results, as presented in the table, were obtained for the E-field probe in which different combinations of the same detection network were applied while filter capacitances C_f and antennas were replaced. For the calculations and measurements, the antennas of following sizes were applied:

- A symmetrical dipole antenna h = 5 cm, 2a = 1.0 mm
- A capacitive extended antenna h = 5 cm, h_l = 3.5 cm printed on a substrate in the form of 5 mm wide strips
- A biconical antenna printed on a substrate in the form of two triangles where h = 5 cm and the triangle base = 3.5 cm

The comparison allows us to formulate several comments:

1. The measured magnitudes of transmittance are below those estimated, which is the result of using formulas that are valid for spatial structures for the flat structures parameters estimation
2. For similar reasons there is a discrepancy between the calculated and the measured magnitudes of the quality factor and the gain
3. The measured lower corner frequencies are, in general, below those calculated, and it allows us to suppose that the estimated equivalent resistance of the detector was underrated
4. The quality factor and the gain of the probe with the capacitive extended dipole antenna and those of the biconical one are similar, however, because of more soft input impedance variations of the biconical antenna versus frequency, the antenna may be accepted as more convenient for the super wideband measurements
5. The presented method of the E-field probes parameters' comparison may also be applied as a comparison criterion in aspects other than the probe's antenna type, for instance, for comparison and optimization of other designs of probes and meters.

Antenna	C_f [pF]	Theory T_0 mV/V/m	Theory f_1 [kHz]	Theory q	Experiment T_0 mV/V/m	Experiment F_1 [kHz]	Experiment q	Gain g Theory	Gain g Experiment
Thin Symmetrical Dipole	0	12.0	34.6		11.6	21.2	5.47		
	10	2.24	6.47	3.45	2.20	3.8	5.79		
	20	1.23	3.57		1.21	2.11	5.73		
Dipole with Capacitive Extension	0	38.5	20.6		25.2	14.5	17.4		
	10	10.7	5.74	18.7	5.51	3.52	15.7	5.4	2.9
	20	6.20	3.33		3.13	1.18	16.6		
Biconical	0	37.3	20.2		26.5	15.1	17.5		
	10	10.6	5.71	18.5	5.81	3.48	16.7	5.4	3.2
	20	6.15	3.33		3.33	1.89	17.6		

Table 9. *Comparison of parameters of E-field probes with different antennas.*

4.6. Comments and conclusions

The E-field probe comparison has allowed us to formulate several conclusions regarding the optimization of antenna types used in them. The conclusions are, in some sense, evident but the most important are the approaches that may be used by meter builders when the probes are planned and designed as well as their users for evaluation and comparison of meters available on the market.

However, the most important thing is the possibility of estimating the measurement accuracy that is a function of all accuracy limiting factors discussed in this chapter. In addition, other factors that are independent of the particular needs of a measurement, should be found, defined, estimated and included in the resultant evaluation of the measurement inaccuracy by the reader.

There is a possibility of almost arbitrary shaping of the frequency response of a wideband probe by the appropriate choice of the corner frequencies of the filters used in shaping the response. For instance, based on the permissible magnitudes of errors δ_{1E} and δ_{4E}, it allows remarkable limitation of the probe's sensitivity at frequencies outside the measuring band. On the other hand, the errors resulting from the accepted relations of R_0/h and h/λ determine both the properties of the probe with a selected type of the measuring antenna and the applicability of the probe in the near-field measurements. Consequently, we devoted most of our attention to their analysis.

The errors δ_{1E}, δ_{2E}, δ_{3E}, δ_{4E} and δ_{5E} were estimated only for symmetrical dipole antennas. The uselessness of unsymmetrical antennas for near-field EMF measurements may be shown based on the capacitive couplings of the antenna with surrounding objects and the role of the person taking the measurements as a 'counterpoise' of the antenna. Presentation of other types of antennas, for instance: capacitive extended dipoles, cylindrical or biconical ones, was omitted because, apart from more precise quantitative relations, as the qualitative ones are identical, it would not bring anything new as compared to the presented considerations.

The presented considerations allow precise estimation of the maximal magnitudes of measuring errors resulting from the size of the probe's antenna as well as from the distance between the antenna and the radiation source and, as a result, allows us estimate the maximal sizes of the measuring antenna. Inversely, it allows us to establish the minimal distance from a source at which the measurement will be loaded by an error of acceptable magnitude. It is especially important when studying the market's offering because the information on the measuring error increase when the measurements are performed close to material bodies, is not published by the meters' manufacturers. This is because the boosting purposes instead reveal much smaller errors valid only in the far-field.

The errors resulting from alternations of the measured field phase along the antenna are identical for the longitudinal and the transversal field components, whereas errors resulting from R_0/h ratio are maximal while the longitudinal component is of concern. Moreover, the transversal component of the electric field approaches zero near a perfectly conducting surface. Because of this, errors δ_{2E} and δ_{5E} were calculated for the longitudinal component, but during the component measurement the antenna is placed perpendicularly to the surface of the source.

The influence of the conducting surface was analyzed when the antenna is located parallel to it (error δ_{3E}). Thus, the error is overestimated when the longitudinal component is measured; as a result it maximizes all possible errors resulting from the effect. The approach, however, was necessary as the measurements may be performed near complex radiating structures, containing a number of active and passive radiators, and was done in order to show the domination of errors δ_{2E} and δ_{5E}. The role of these errors may be demonstrated in the calibration conditions in a TEM cell. When the probe is brought nearer to one of the cell walls the meter indications increase.

The errors δ_{1E} and δ_{4E} are the systematic ones and they may be precisely calculated for every measuring antenna if the frequency of the source is known. Errors δ_{2E}, δ_{3E} and δ_{5E} are systematic ones as well and they may be estimated when the measurement conditions are known (known parameters of the source, measuring antenna and the geometry of propagation). Usually however, the parameters of the source, and as a result the propagation geometry, may not be known beforehand, sometimes even the source itself may not be known before the measurements. As mentioned above, the probes' (meters') parameter are not revealed by their manufacturers. It leads to the necessity of assuming (especially when measurements of a surveying or monitoring character are performed) that these errors are accidental ones.

It may be assumed that the sizes of the measuring antenna are usually much smaller in comparison to those of the radiation source. Thus, the source may be considered, in relation to the probe, as 'an infinitely long line' generating in its vicinity the cylindrical wave of which the radial E-field component dominates in its proximity. In the case where the error does not exceed its magnitude estimated for $\alpha = 1$, it may be recognized as the most probable error when the field sources of practical importance are investigated.

We should notice that the errors δ_{2E} and δ_{3E} are always positive; which causes the result of measurement near material media to be overestimated in relation to the standardized conditions. It explains the increase of the meter's readouts while measurements are performed close to a primary or a secondary source. From a purely metrological point of view the sign of an accidental error is of relatively low importance. However, in our field, where the possible EM radiation hazard is of concern, positive character of the error may be, in some sense, an advantage as the real field strength should not exceed the measured one. However, the statement by no means should change our trend to measure the E-field (and any other physical quantity) with the desired and possible accuracy estimate.

4.7. Bibliography

1. T. Babij, H. Trzaska, "Wideband Properties of Electric Field Probes," *Proc. 1975 IEEE Int'l. EMC Symp.*, pp. 5BIa1-6, San Antonio, TX.
2. R. W. P. King, *The Theory of Linear Antennas*, Harvard Univ. Press, 1956.
3. M. Kanda, "Analytical and Numerical Techniques for Analyzing an Electrically Short Dipole with a Nonlinear Load," *IEEE Trans.*, Vol. AP-48, No. 1/1980, pp. 437–442.
4. G. Z. Ajzenberg, *Shortwave Antennas* (in Russian), Sviazizdat, Moscow, 1962.

5

Magnetic Field Measurement

This chapter discusses probe properties for RF magnetic field measurements and in particular, the factors limiting measurement accuracy. Our consideration is limited to a probe consisting of a circular loop antenna loaded with a detector of a shaped frequency response. The majority of the presented estimations are fully applicable for Hall-cell probes, magneto-optic probes, magneto-diode probes, and for other designs, especially when considering the averaging of the measured field upon the surface of the measuring antenna (probe).

Parts of our considerations are similar in character to those presented in Chapter 4, and parts are concerned with the magnetic field measurement specificity.

5.1. Measuring antenna size

In the analogy used in Chapter 4.1, it is possible to follow the discussion relating to the measuring antenna size limitation resulting from the error of a quasi-point value of the magnetic field measurement. It concerns the electrical sizes of the antenna in particular. The discussion leads to conclusions similar to those presented in chapter 4.1 if we use πr (where r = radius of a circular loop antenna) instead of h. However, a more rigorous limitation of the loop antenna size results from the so-called "antenna effect," in other words, from simultaneous sensitivity of the loop antenna (especially a non-screened one) to the electric field.

The current in a load of the loop antenna I_l may be expressed as:

$$I_l = \lambda K_H cB \pm \lambda K_E E = I_H \pm I_E \qquad (5.1)$$

where K_H and K_E = sensitivity of an unloaded loop to the magnetic and electric field respectively, and c = in general, the velocity of propagation, here the velocity of propagation in a vacuum:

$$c = \frac{1}{\sqrt{\varepsilon_0 \mu_0}}$$

and I_H and I_E = components of the current in the antenna load proportional to the magnetic and the electric field.

Formula (5.1) was introduced for the plane wave and positioning of the loop in relation to the field components such that I_H is of maximal value while I_E is of maximal or minimal value. The resulting measurement error due to the presence of the antenna effect we will define as:

$$\delta_{1H} = \frac{|I_H - I_E|}{|I_H|} \approx \frac{|K_E|}{|K_H|} \frac{|Z_a|}{Z_0} \qquad (5.2)$$

where Z_a = wave impedance:

$$Z_a = \frac{E}{H} \qquad (5.3)$$

independent from the source type $|Z_a| \in <0, \infty>$, Z_0 = intrinsic impedance of free space:

$$Z_0 = \sqrt{\frac{\varepsilon_0}{\mu_0}} \approx 120\pi \ [\Omega]$$

E, H = complex amplitudes of the E and H vectors.
In accordance with King [1]:

$$\frac{K_E}{K_H} = -j2\pi \frac{2r_0}{\lambda} \qquad (5.4)$$

To reduce the significance of the antenna effect, a symmetrization and/or screening of the loop is applied. A double- or multiple-loaded loop can serve the same purpose. An example of a multiple-loaded loop application will be presented in Chapter 6. The error δ_{1H} response as a function of kr_0 for several values of Z_a is plotted in Figure 5.1.

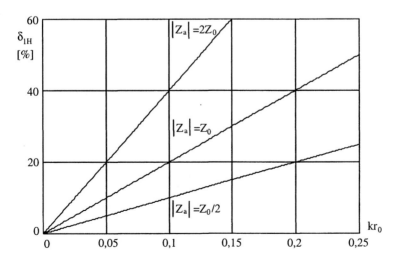

Figure 5.1. *The error δ_{1H} response as a function of kr_0.*

Formulas (2.21), (2.22) and (2.23) as well as (5.2) allow us to write the elemental magnetic dipole as:

$$\lim_{R_0 \to 0} \delta_{1H} \to \infty \qquad (5.5)$$

In light of this, the magnetic field measurement close to a short electric antenna (where the electric field is the "dominant" one) may lose its sense. A radical way of reducing the antenna effect is to use an electrostatic screen. However, this is not an ideal solution because the simultaneous increase in the antenna capacitance results in a decrease of both the antenna's resonant frequency and the sensitivity of the probe with such an antenna. Analogous to the considerations in Chapter 4.3, and Formulas (2.28), (2.29) and (2.30), it is possible to say, for an infinitely long line source, if: $z \to 0$ and $\rho \to 0$ then $|Z_a| \leq Z_0$. However, for $z \to h$, the inequality should be reversed. In questionable cases, as a test of the measurement error magnitude resulting from the antenna effect, we should base our investigation on the asymmetry, σ, of the directional pattern of the measuring antenna (probe). The criterion is simple and may be used for both for single loop probes and omnidirectional ones:

$$\sigma = \frac{\alpha_1 - \alpha_2}{\alpha_1 + \alpha_2} = |\delta_{1H}|_{max} \qquad (5.6)$$

where α_1 and α_2 = indications of the magnetic field meter responding to both maxima of its directional pattern.

5.2. Frequency response of the magnetic field probe

The effective length, h_{eff}, of a loop antenna fulfilling the condition $kr_0 << 1$ is direct proportional to the frequency. In order to obtain frequency independent response of the magnetic field probe, while the probe works at frequencies below the resonant

MAGNETIC FIELD MEASUREMENTS 91

frequency of the antenna, the antenna is loaded with a four-terminal network whose frequency response is proportional to λ.

An equivalent network of the magnetic field probe at low frequencies, comprised of an antenna and a response-correcting network, is shown in Figure 5.2.

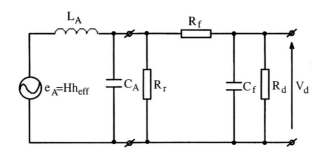

Figure 5.2. *Equivalent network of the magnetic field probe.*

The transmittance of the probe is given by (5.7):

$$T(jf) = \frac{V_d}{H} = \frac{-2\pi m S \mu_0 f}{\left(1 - \frac{f^2}{f_1^2}\right)\left(1 + \frac{f_2}{f_3}\right) - \frac{f^2}{f_1^2}\frac{f_4}{f_3} + j\left[\left(1 - \frac{f^2}{f_1^2}\right)\frac{f}{f_3} + \frac{ff_4}{f_1^2}\left(1 + \frac{f_2}{f_3}\right)\right]}$$

where

$$f_1 = \frac{1}{2\pi\sqrt{L_A C_A}}$$

$$f_2 = \frac{1}{2\pi R_d C_f}$$

$$f_3 = \frac{1}{2\pi R_f C_f}$$

$$f_4 = \frac{1}{2\pi R_r C_A}$$

92 ELECTROMAGNETIC FIELD MEASUREMENTS IN THE NEAR FIELD

$$h_{eff} = 2\pi \, nS\mu_0 f \tag{5.8}$$

n = number of turns,
S = the surface area of the loop antenna.

To achieve an optimal shape of the probe's frequency response it is convenient to assume:

$$R_r = 2\pi f_1 L_A$$

In other words, $f_1 = f_4$

On the grounds of Formula (5.7), it is possible to determine the following frequency segments of the magnetic field probe:

a) At low frequencies:

$$|T(jf)| = \frac{2\pi n S \mu_0 R_f f}{R_f + R_d} \tag{5.9}$$

b) At medium frequencies for which $f_2 \ll f_3$ is always fulfilled, $T_0(jf) \neq f(f)$ and the transmittance is given by:

$$T_0(jf) = \frac{2\pi n S \mu_0}{R_f C_f} \tag{5.10}$$

Note that the lower corner frequency of the frequency band, in which the transmittance decreases by 3 dB, $f_l = f_3$, whereas the upper corner frequency of the probe $f_u \approx 1,5 \, f_1$. However, it is usually applied $f_4 \approx 2f_1$ which means that the transmittance has a certain maximum for $f = f_1$. Sometimes, in order to eliminate the maximum, another RC low-pass filter or two of $f_g \approx 0.5 f_1$ is applied, then $f_u = f_g$.

c) High frequencies at which:

$$|T(jf)| = \frac{2\pi n S\mu_0 f_1^2 f_3}{f^2} \qquad (5.11)$$

The calculated run of the magnetic field probe's frequency response, whose equivalent diagram is shown in Figure 5.2, for $f_4 = 2f_1$ is shown in Figure 5.3 (continuous line). The dashed line generalizes the measured frequency response of the probe equipped with an additional double RC low-pass filter of $f_g = 0.5f_1$. Both curves were normalized in relation to T_0.

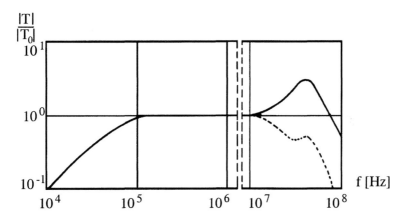

Figure 5.3. *Transmittance of the wideband magnetic field probe.*

An accurate analysis of the magnetic field probe equipped with a triple RC low-pass filter has been performed. The first segment of the filter allows us to obtain the required lower corner frequency and the two-segment filter to lets us attenuate signals at the antenna's self-resonant frequency, eliminate any other maxima, and reduce the probe's sensitivity above its upper corner frequency. Especially important is the frequency at which there is maximum attenuation and where the antenna still works as a "folded dipole" with relatively low input impedance. Our analysis takes into account the HF equivalent network. Even though we found that the local maxima of transmittance may appear at frequencies corresponding to the resonance of the antenna's self inductance and the parasitic reactance of the probe, the magnitude of the transmittance at these maxima does not exceed −20 dB in relation to the transmittance within the measuring band. Our estimations and experiments have confirmed that the use of additional RC filters is indispensable. Without them, above about 100 MHz, the transmittance of a probe, designed to work up to 20-30 MHz may be out of control. While fringes at these frequencies appear during measurements they may lead to errors totally falsifying the results. It is important to be able to control the frequency response run of any probe outside the measuring band, and especially at the highest frequencies. In addition, the person performing the measurements should know the frequency response run.

5.3. Directional pattern alternations

The solution of the integral equation, that describes the current intensity along the circular loop antenna shown in Figure 5.4, may be presented in the form [2]:

$$I(\varphi) = \frac{-jV_0}{Z_0 \pi} \left(\frac{1}{a_0} + 2 \sum_{n=1}^{\infty} \frac{\cos n\varphi}{a_n} \right) \qquad (5.12)$$

where φ = the central angle of the antenna
V_0 = the voltage of the antenna exciting source
a_n = a coefficient of the Fourier series expansion.

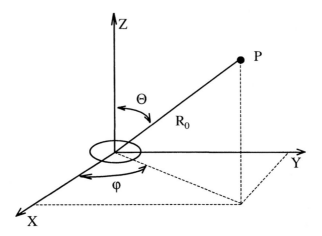

Figure 5.4. *A circular loop in a coordinate system.*

After substitution of formula (5.12) in to formula (2.8) and then in to formulas (2.13) and (2.14), we solve for the spatial components of the magnetic field in the far-field. For a loop antenna, located in the coordinate system as shown in Figure 5.4, which fulfills the condition $kr_0 \leq 0.2$, the dominant component is $H_{\theta 0}$ (where index zero indicates zeroth order of approximation). This component corresponds to using only the term a_0 in the series (5.12) which means that the uniform current distribution along the antenna (characteristic to the elemental dipole), is then:

$$[H_{\theta 0}]_{xy} = \frac{j\omega\varepsilon_0 V_0 r_0}{2R_0} J_1(kr_0 \sin\theta) = AJ_1(kr_0 \sin\theta) \quad (5.13)$$

If, in the series (5.12) two terms are considered, i.e., a_0 and a_1, then the approach is equal to the first order of approximation that is equivalent to the cosine current distribution. The approximation is true for $kr_0 \leq 1$, then:

$$[H_{\theta 1}]_{xy} = A_1 J_1(kr_0 \sin\theta) \cos\varphi \quad (5.14)$$

$$[H_{\varphi 1}]_{xy} = A_1 \frac{J_0(kr_0 \sin\theta)}{kr_0 \sin\theta} \cos\theta \sin\varphi \qquad (5.15)$$

where A and A_1 = constants dependent on R_0
J_1 = Bessel function of the first order.

If three identical loop antennas are placed in three mutually perpendicular planes and their directional patterns are summed with square, that is equivalent to the load of each antenna with a square-law detector, then we obtain a magnetic field probe with a spherical radiation pattern (omnidirectional probe). It is shown below. As a reminder, that although the radiation pattern is calculated in specific way for a transmitting antenna then, by using the reciprocity theorem, the results are applied for a receiving antenna. However, such an antenna may only work only as a receiving one.

Let's consider the radiation pattern of a system consisting of three mutually perpendicular loop antennas for the zeroth order approximation of their current distribution. If the radiation pattern f_{xy} of a loop antenna is placed upon the xy plane of a Cartesian coordinate system, we define it in the form:

$$f_{xy} = \frac{[H_{\theta 0}]_{xy}}{[H_{\theta 0}]_{xy\,max}} = \frac{J_1(kr_0 \sin\theta)}{kr_0} \qquad (5.16)$$

And those of the antennas on the planes xz and yz in the form:

$$f_{xz} = \frac{J_1(kr_0\sqrt{1 - \sin^2\theta \sin^2\varphi})}{kr_0} \qquad (5.17)$$

and

$$f_{yz} = \frac{J_1(kr_0\sqrt{1 + \sin^2\theta \cos^2\varphi})}{kr_0} \qquad (5.18)$$

Thus, for x << 1 and $J_1(x) \approx x$, and by analogy to formula (4.42), for antennas loaded with the square-law detectors, we have:

$$F(\theta,\varphi) = f_{xy}^2 + f_{xz}^2 + f_{yz}^2 = 2 \qquad (5.19)$$

The directional pattern given by formula (5.19) is direction independent with no regard to the applied coordinates. It is true for $kr_0 \ll 1$. For our purposes this is an extremely interesting case, as it results from the considerations presented in chapter 5.1.

For antennas that do not fulfil the condition $kr_0 \ll 1$, it would be necessary to repeat the estimations, for instance, for the first approximation, for which the radiated field is given by formulas (5.14) and (5.15) and to accept a definition of the pattern irregularities identical to that given by formula (4.43). If an antenna fulfils the condition $2\pi r_0 \leq \lambda$, the results of the pattern's non-uniformity estimations are identical to those presented in Figure 4.4, when the substitution $h = \pi r_0$ is used.

As in chapter 4.4, during estimations of the spherical directional pattern of a dipole antenna, when the spherical pattern of three mutually perpendicular loop antennas was synthesized, the amplitude variations on the surface of the antennas were found to be the most influential. It was shown that if $kr_0 \leq 0.2$, the role of the phase variation upon the shape of the pattern is negligible. Thus, the sizes of the loop antenna in relation to the wavelength are limited by the permissible magnitude of the error δ_{1H}. Again, per analogy to the pattern deformation δ_{5E} (formula 4.48), it is possible to estimate the irregularities of the spherical pattern of a system of three loop antennas as a function of r_0/R_0. The estimation may conclude that the non-uniformity is approximately twice as much as the error δ_{2H}. Similar conclusion has already been formulated for the electric field probe.

5.4. Accuracy of measurement versus distance of the antenna to the source of radiation

5.4.1. The influence of the source structure upon accuracy

The electromotive force e_H induced by the field in a loop antenna of surface area S is given by Faraday's law:

$$e_H = -j2\pi n\mu_0 f \int_S \mathbf{H} \cdot d\mathbf{S} \qquad (5.20)$$

If the source of the radiation is the magnetic elemental dipole located at the point P (R_0, q, φ) at distance R_0 from the center of the circular loop antenna, as shown in Figure 5.4, and the sizes of the dipole are much less than R_0, then, making the use of formula (2.27), we write:

$$e_H = -j2\pi n\mu_0 fC_2(\alpha) \int_S \frac{\exp(-jkR)}{R^\alpha} \cos(\mathbf{H}, d\mathbf{S}) \, dS \qquad (5.21)$$

where R = distance between the integration point and the field source:

$$R = \sqrt{R_0^2 + x^2 + y^2 - R_0 \sin\theta \, (x\cos\varphi + y\sin\varphi)} \qquad (5.22)$$

and x, y = coordinates of the integration point: x, y ∈ S.

The maximal phase and amplitude shifts of the field on the surface of an antenna will appear for the transversal field component, when **H** is parallel to **S** and q = $\pi/2$. If we assume, for instance, $\varphi = 0$ (which is not essential because of symmetry) we have:

$$e_H = -j\frac{h_{eff}}{S} C_2(\alpha) \int_S \frac{\exp(-jkR)}{R^\alpha} \, dxdy \qquad (5.23)$$

and

$$R = \sqrt{(R_0 - x)^2 + y^2} \qquad (5.24)$$

where h_{eff} = effective length given by formula (5.8).

The presence of the antenna effect limits the loop antenna size more rigorously, as compared to field averaging over the surface of the antenna caused by the phase differences on the surface. In the same manner as Chapter 4.2, it allows the assumption in formula (5.23) that $\exp(-jkR) \approx \exp(-jkR_0)$, then:

$$e_H = -j \frac{h_{eff}}{S} C_2(\alpha) \exp(-jkR_0) \int_S \frac{1}{R^\alpha} dxdy \qquad (5.25)$$

In the case of the homogeneous field, the emf e'_H is:

$$e'_H = -h_{eff} C_2(\alpha) \exp(-jkR_0) R_0^{-\alpha} \qquad (5.26)$$

In order to introduce a uniform reference system (standardization) it is indispensable to assume, as presented elsewhere in this book, that the probe's calibration is performed in the homogeneous field whose curvature radius is equal to infinity. As a result, the error δ_{2H} will appear, which results from averaging the field of finite curvature upon the surface of the probe when magnetic field measurements are performed in close proximity to a source. We define the error in the form:

$$\delta_{2H} = \frac{e_H - e'_H}{e_H} \qquad (5.27)$$

After substituting formulas (5.25) and (5.26) in to formula (5.27) we have:

For α = 3

$$\delta_{2H} = 1 - \left\{ \frac{4}{\pi k^2} \left[\frac{1}{1-k^2} E(k) - K(k) \right] \right\}^{-1} \qquad (5.28)$$

where E(k) and K(k) are elliptic integrals of the first and second order and argument k:

$$k = r_0/R_0$$

For α = 2

$$\delta_{2H} = 1 - k^2 \left[\ln \frac{1}{1-k^2} \right]^{-1} \qquad (5.29)$$

where $k = r_0/R_0$, and for α = 1

$$\delta_{2H} = 1 - \frac{\pi r_0}{2R_0} \left[\left(\frac{R_0}{r_0} + 1 \right) E(k) - \left(\frac{R_0}{r_0} - 1 \right) K(k) \right]^{-1} \qquad (5.30)$$

where E(k) and K(k) are elliptic integrals of the first and second order and argument k:

$$k = \frac{2\sqrt{R_0 r_0}}{R_0 + r_0}$$

and for α = 0, $\delta_{2H} \equiv 0$.

Calculated values of the error δ_{2H} are plotted in Figure 5.5.

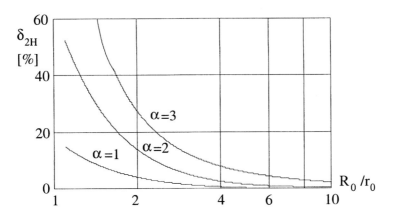

Figure 5.5. *Error δ_{2H} as a function of R_0/r_0.*

5.4.2. The role of antenna input impedance alternations

The transmittance of the magnetic field probe, described in chapter 5.2, is independent of the input impedance of the probe's antenna, which is derived from formula (5.10). As a result, consideration of the antenna's input impedance changes, in proximity to a conducting medium, could be disregarded. However, attention should be focused on the fact that the overall measurement error, resulting from the measured field averaging (especially in the near-field), which is a function of r_0/R_0 for the considered probe design, is only a function of the structure of the radiation source (Figure 5.5) and that the character of the error is always positive.

In Chapter 5.5 related slightly differently when compared to Figure 5.2, we will briefly show how estimation of the loop antenna input impedance varies as a function of distance when compared to a conducting medium. The medium, as before, is infinitely large and conducts perfectly. The presence of such a medium may be replaced by a mirror reflection of the antenna under consideration.

Mutual inductance, M, of two identical loop antennas, collinearly placed at distance $2R_0$, is given by the ratio of the magnetic flux of the first antenna flowing thorough the surface area of the other, S_2, to the current I_1 which the flux generates [3]:

$$M = \frac{\int_{S_2} B_1 dS_2}{I_1} \qquad (5.31)$$

After integration it becomes:

$$M = \mu_0 r_0 k \left[\left(\frac{2}{k} - k \right) K(k) - \frac{2}{k} E(k) \right] \qquad (5.32)$$

where E(k) and K(k) are elliptic integrals of the first and second order and argument k:

$$k = \left[\left(\frac{R_0}{r_0} \right)^2 + 1 \right]^{-2}$$

For $R_0 \to 0$ and $M \to L_A$, then:

$$L_A = \mu_0 (2r_0 - a) \left[\left(1 - \frac{k^2}{2} \right) K(k) - E(k) \right] \qquad (5.33)$$

Where E(k) and K(k) are elliptic integrals of the first and second order and argument k:

$$k = \frac{2\sqrt{r_0(r_0 - a)}}{2r_0 - a}$$

2a = diameter of the conductor applied for the antenna windings; if $a/r_0 \to 0$, then:

$$L_A = \mu_0 r_0 \left[\ln\left(\frac{8r_0}{a}\right) - 2 \right] \qquad (5.34)$$

Other approaches may yield similar results [4]. For the first approximation of the current distribution along the antenna, using formula (5.12), the input impedance of the antenna Z_A may be given in the form:

$$Z_A = \frac{j\pi Z_0 a_0 a_1}{2a_0 + a_1} \qquad (5.35)$$

If we assume beforehand that it is possible to construct a magnetic field probe whose transmittance within the measuring band is given by:

$$T(jf) = \frac{A}{L_A} \qquad (5.36)$$

where A = constant

Then the measurement error, caused by variations of the probe's (antenna) input impedance changes due to its parallel location at distance R_0 from a conducting plate as defined above, will be:

$$\delta_{3H} = \frac{M}{L_A} \qquad (5.37)$$

The calculated error values, δ_{3H}, in the function of R_0/r_0, for several r_0/a ratios are shown in Figure 5.6. Figures 5.5 and 5.6 show that the errors δ_{2H} and δ_{3H} are of positive character, thus, the resultant error is also positive. The error gives a non-understated value of measured field strength and may be, in some sense, advantageous when labor safety or environment protection is of concern.

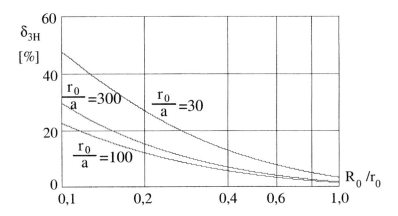

Figure 5.6. *Error δ_{3H} as a function of R_0/r_0.*

5.5. The magnetic field probe with a loop working above its self-resonant frequency

In Figure 5.1, we showed an equivalent circuit of the magnetic field probe in which, to achieve the frequency independent transmittance within a certain frequency band, the loop antenna was followed by an RC lowpass filter. The solution is inconvenient, especially when detectors of low input impedance, i.e., thermistors or thermocouples, are to be used. The latter has several advantages (more univocal square-law dynamic characteristics, less susceptibility to interference, immune to static electricity).

We proposed using the properties of electrically small loop antennas working at frequencies above self resonance as an equivalent alternative to the magnetic field probe discussed in Chapter 5.2. To clarify, the antenna has many resonant frequencies, but here we are considering the resonance of the small size antenna inductance and parasitic capacitance of the probe, i.e., self-capacitance of the antenna eventually increased by an electrostatic screen, input capacitance of the antenna load and the capacitance of the probe assembling. While the properties of such a probe are being considered we will use a formula defining the effective length of the loop antenna with an approximation better than that of zeroth order (formula 5.8). Based on formula (5.12), and using the reciprocity theorem, it is possible to estimate the effective length. Taking into account the phase distribution on the surface S of the antenna, its effective length may be expressed in the form:

$$h_{eff} = 2\pi n \mu_0 f S \left[1 - 2^{-3}(kr_0)^2 + \frac{2^{-6}}{3}(kr_0)^4 + \cdots \right] \qquad (5.38)$$

Notice that the zeroth approximation of the loop antenna effective length given by formula (5.8) corresponds to the first term of the series in formula (5.38). The transmittance of the magnetic field probe, composed of a loop antenna loaded with a serial connected RC filter (Figure 5.7), we define as the ratio of the current, I, flowing in the loading resistance (and in the antenna), R_S, to the intensity of the measured magnetic field H:

$$T(jf) = \frac{I}{H} = \frac{h_{eff}}{Z_s} \qquad (5.39)$$

where Z_S = the equivalent impedance of the serial connection of the antenna input impedance Z_A and R_s and C_s.

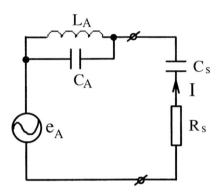

Figure 5.7. *Equivalent network for the medium frequency range of an H-field probe.*

Formula (5.39) is convenient when considering the properties of this type of probe. However, in order to obtain results that are comparable to those obtained from formula (5.7), it is necessary to multiply formula (5.36) by R_s. Then substituting h_{eff}, given by formula (5.38), and Z_A, given by (5.35) in to formula (5.39), we have for the medium frequency band:

$$T_0(jf) = \frac{jnS\mu_0 R_s}{L_A} \qquad (5.40)$$

We might notice that formula (5.40) is analogous to formula (5.36).
For a maximally flat frequency response, in other words, for:

$$R_s = \sqrt{L_A/C_s}$$

the corner frequencies (−3 dB) of the medium frequency band are:

a) The lower corner frequency:

$$f_1 = 0{,}62\, f_2 \qquad (5.41)$$

b) The upper corner frequency:

$$f_u = 0{,}5\, f_1 \qquad (5.42)$$

where

$$f_1 = \frac{1}{2\pi\sqrt{L_A C_A}} \qquad (5.43)$$

$$f_2 = \frac{1}{2\pi\sqrt{L_A(C_A + C_s)}} \approx \frac{1}{2\pi\sqrt{L_A C_s}} \qquad (5.44)$$

$$f_3 = \frac{1}{2\pi R_s C_s} \qquad (5.45)$$

In this estimation, the above-mentioned effects, resulting from the resonance of the antenna as a folded dipole, have not been taken into account. As in the case of the antenna working below its self-resonant frequency, we require the use of additional protection against unwanted increases in the probe's sensitivity at frequencies above the measuring band. Figure 5.8 shows plotted results for simplified calculations of the transmittance of a magnetic field probe while $f_3/f_1 = 1$, 3 and 10. The calculations made it possible to optimize the run of the transmittance in the low frequency range. On the other hand, Figure 5.9 shows the frequency response of the probe measured for $f_3 = f_1$. An uncontrolled increase of the probe's sensitivity should be noticed in the probe's frequency response without an electrostatic screen. This can be explained as the approach to the frequency range in which the mentioned dipole resonances should be taken into consideration. The curves presented in Figures 5.8 and 5.9 were normalized in relation to T_0.

108 ELECTROMAGNETIC FIELD MEASUREMENTS IN THE NEAR FIELD

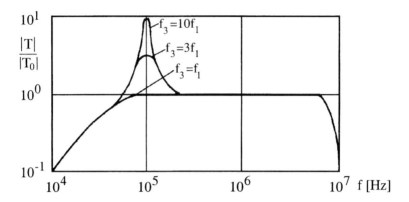

Figure 5.8. *Estimated frequency response of the magnetic field probe.*

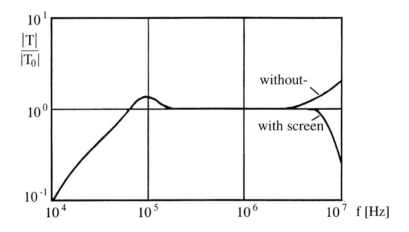

Figure 5.9. *Measured frequency responses of the H-field probe.*

This approach was introduced to help us evaluate the applicability of a selected type of magnetic field probe, and with consideration of the electric field probes presented in Chapter 4, the measure of the probe's quality defined in the form:

$$q = T_0(jf) \frac{f_u}{f_l} \qquad (5.46)$$

By calculating the quality factor, q, and, as before, the defined gain factor of the magnetic field probes presented in Chapters 5.2 and 5.5, it is possible to show that, while loop antennas of similar parameters are used in the probe working at frequencies above and below the antenna self-resonance, the parameters of both probes are equivalent to one another.

5.6. Comments and conclusions

The maximum size of the magnetic field probe, in which the loop antenna serves as the field sensor, for which an acceptable magnitude of the error δ_{1H}, may be estimated with the use of curves is shown in Figure 5.1. However, it may be necessary to use probes with different diameter antennas for measurements performed near different sources of radiation. This is because the error is a function of E/H and the latter characterizes the sources. This is an important possibility especially when known field sources are being investigated. Unfortunately, in the majority of cases, the source's identification may be difficult, if not impossible. This suggests that we take extra care and make an assumption that we will have the maximal errors. For the purpose of determining the magnitude of error δ_{1H} and its importance, evaluation of the criterion given by formula (5.6) may be useful. As mentioned above, the loop antenna size limitation resulting from the presence of the antenna effect is much more rigorous as compared to that resulting from the phase averaging on the antenna surface.

The permissible magnitude of the error δ_{1H} is a starting point for evaluation of the upper corner frequency and the design of the filters for shaping the detector's frequency response. With no regard to the assumed concept of the probe's work (below or above the self-resonance), the upper corner frequencies for identical antennas are similar. When antenna sizes are known, it is possible to estimate the errors δ_{2H} and δ_{3H} as functions of R_0/r_0. For the above-resonance probe, it is necessary to take into account both errors whereas for the below-resonance probe, the variations of the antenna's input impedance do not have a remarkable influence upon the measurement accuracy.

Similar to the considerations presented in Chapter 4, and for the error estimations in Chapter 5, we have assumed the most disadvantageous measurement conditions that lead to the maximal error calculations. In order to find the maximal value of the error δ_{2H}, the error was calculated for a situation equivalent to the longitudinal field component. For example, the situation takes place near a heating coil of an induction furnace where the spatial field variation allows the assumption a = 3. However, if the magnetic fields associated with high frequency currents in an 'infinitely long cable' are measured, the calculation gives much more realistic results even for a = 1. It results from the occurrence and domination of the tangential field component near the surface of the conductor. This is an important difference between the electric and the magnetic field and, as a result, the accuracy difference of both components' measurements.

The error δ_{3H} was calculated for parallel positioning of the antenna in relation to a conducting surface. As a result, during the transverse component of the field measurement, the value of the error is overestimated and it maximizes all possible error values. Consequently, this approach is applied to the presented example and it is supported by the possibility of performing measurements near primary and/or secondary sources of complex and unknown structure.

Because of an indeterminate field source, which occurs in applications specific to our considerations, we will assume that the errors δ_{2H} and δ_{3H} are accidental. Simultaneously, the use of small size antennas (in the electric and geometric sense) allows the supposition that the most probable error, δ_{2H}, is that estimated for $\alpha = 1$.

Based on the presented analyses, it is possible to attempt an estimation of the resultant error of the magnetic field measurement using one of the probe's presented versions or using another probe's design (e.g., a probe with the Hall-cell). The situation may be obvious and simple enough that precise determination of any discussed partial errors is analytically possible. Moreover, this creates the temptation to increase the accuracy of the field measurement via the individual measurement accuracy estimation or, at least, making a use of analytically determined correction factors while final results of measurements are completed. To repeat it again: the procedure is acceptable and fully possible only in the case where the measurement conditions are fully known. In the metrological practice these circumstances may appear rather incidentally; thus, the approach should be allowed only in a very limited number of cases. However, assuming that the sizes of antennas (probes) applied correspond to the limitations given by error δ_{1H}, with inaccuracy to the error, it may be said that the accuracy of the magnetic field measurement is limited mainly by the magnitude of error δ_{2H}. In other words, the error results from the amplitude of measured field averaging along (on the surface) of the measuring antenna. At the same time the most probable magnitude of error corresponds to the assumption $a = 1$. The identical conclusion has already been formulated at the end of Chapter 4.

It is worth our while to remember this newly proposed method of magnetic field measurement based upon the current in the antenna measurement using a current transducer. This method allows a remarkable sensitivity increase and stability of measurement, however, it is at the expense of the frequency response uniformity [5]. The intensity of the current flowing in a loop antenna is directly proportional to the magnetic field strength and the proportion is valid through a wide frequency range that allows considerable simplification of a meter designed in accordance to the concept. The frequency response non-uniformity is affected mainly by the imperfection of the applied transducers. It is this author's hope that after their development, the method may be dominant in the future, especially at higher frequencies.

5.7. Bibliography

1. H. Whiteside, R. W. P. King, "The Loop Antenna as a Probe," *IEEE Trans.*, Vol. AP-7, No. 5/1964, pp. 291–294.
2. R. W. P. King, "The Loop Antenna for Transmission and Reception," Chapter 11 in: T. E. Collin, F. J. Zucker, *Antenna Theory Part I*, McGraw-Hill, 1969.
3. J. D. Kraus, *Electromagnetics,* 3rd edition, McGraw-Hill, 1984.
4. S. Ramo, J. R. Whinnery, *Fields and Waves in Modern Radio*, John Wiley & Sons. 1953.
5. J. Zurawicki, *EM-Field Measurements* (in Polish). Graduate work at the Technical University of Wroclaw, Poland, 1992.

6

Power Density Measurement

6.1. Power density measurement methods

If electric field strength E, and magnetic strength H, are known, the power density (the power density flux) is explicitly (as to its magnitude, phase and direction) determined by the Poynting vector S. The averaged value of the vector $\mathbf{S_a}$, which expresses the power flow from a source, is the subject of our interest. The quantity is given by:

$$\mathbf{S_a} = \frac{1}{2} \operatorname{Re}(\mathbf{E} \times \mathbf{H}^*) \qquad (6.1)$$

In far-field conditions it is enough to have (calculated, measured) one of the field components as their mutual relationship is known and the relation is expressed by formulas in the form of (2.13) and (2.14). The power density is the sum of the electric power density S_E, and the magnetic one S_H, that may be expressed (changing vector notation to scalar) in the form:

$$S = S_E + S_H = 2S_E = 2S_H \qquad (6.2)$$

In the near-field, the mutual relationship of the E and H fields is unknown beforehand and is a function of the structure of a radiation source as well as the distance between the source and the point of observation. Thus, the power density evaluation based on only one component measurement is flawed because of methodology error. This error is of remarkable interest (and it creates a very important limitation in the use of the method and measuring equipment) because the method is widely applied in measuring devices available on the market. However, nowhere is it explained as to what are acceptable (i.e., to assure required measurement accuracy) conditions of equipment use. We may add here that, although other methods of the S measurement are known and mentioned in other chapters, the method discussed above is the sole method used in the wide spectrum of devices offered on the market.

Before we begin further considerations, we should focus our attention on two problems:

1. We still do not know a method for direct power density measurement (similar to that of the electric and the magnetic component measurements). The measurement requires that we find, as mentioned above, both field components. However, antennas sensitive enough for both components, with one exception that is discussed in chapter 6.2, have resonant sizes and, in light of presented considerations, are useless for near-field metrology. We know from published literature that probe designs exist for power density measurements that fulfil our size requirements and are composed of a number of E- and H-field sensors.

2. With no regard to applied measuring antenna and sensor types, for further consideration we will use only the field relations and the antenna and sensor parameters will not be taken into account. This assumption leads to the determination of the method error. For final estimation of the measurement accuracy it requires that the accuracy of the E- and/or H-field

measurement be included as well (these problems were considered in Chapters 4 and 5).

As before, we accept these assumptions to allow us to find the error values that would maximize measurements errors performed under other conditions. We will estimate errors for such sources as elemental electric and magnetic dipoles. As mentioned in chapter 2, the EMF curvature around these sources is maximal if we take into account sources of practical importance.

Expanding formula (6.1) and substituting formulas (2.21), (2.22) and (2.23) we have for the monochromatic harmonic field an averaged in time, complex magnitude of the Poynting vector:

$$S_m = \frac{1}{2}\left(E_R H_\varphi^* 1_\theta + E_\theta H_\varphi^* 1_R\right) = \frac{k\omega^3 \mu p^2}{32\pi^2 R^2}\left[1 + \frac{j}{(kR)^3}\right]\sin^2\theta \cdot 1_R - \frac{jk\omega^3 \mu p^2}{16\pi^2 R^2}\left[\frac{1}{kR} + \frac{1}{(kR)^3}\right]\sin\theta\cos\theta \cdot 1_\theta \quad (6.3)$$

where 1_R and 1_θ = versors,
p = dipole moment (see formula 2.24).

The power radiated by a source S_r is represented by the real part of formula (6.3). It should be noted that only this part of the vector could be transferred into heat in an absorbing body. This portion of the vector is expressed by a formula that is identical no matter if it is near- or far-field and it is:

$$S_r = \frac{k\omega^3 \mu p^2}{32\pi^2 R^2} \quad (6.4)$$

The magnitude given by formula (6.4) will be used in farther considerations as the reference level for considered methods of measurement.

6.1.1. Power density measurement by E or H measurement

The validity of the power density measurement concept using the E or H field measurement not questioned in the far-field. However, in the near-field, one of the field components may be dominant and, as a result, the power density calculated on the ground of the dominating component measurement will be overestimated and vice versa. Taking into consideration components of the elemental electric dipole, it is easy to see that (for $\theta \to 0$) the following relationship is true:

$$\lim_{R \to 0} |E/H| \to \infty$$

The subject of our measurement is the power density that we will estimate based on the electric field measurement (electric power density) S_E, which for free space conditions and the source in the form of the electric elemental dipole is:

$$S_E = \frac{|E|^2}{2Z_0} = \frac{\omega k^3 p^2}{8\pi^2 \varepsilon R^2} \left\{ \frac{\sin^2 \theta}{4} \left[1 - \frac{1}{(kR)^2} + \frac{1}{(kR)^4} \right] + \cos^2 \theta \left[\frac{1}{(kR)^2} + \frac{1}{(kR)^4} \right] \right\} \quad (6.5)$$

Or, based on the magnetic field measurement (magnetic power density) S_H:

$$S_H = \frac{|H|^2 Z_0}{2} = \frac{\omega k^3 p^2}{32\pi^2 \varepsilon R^2} \left[1 + \frac{1}{(kR)^2} \right] \sin^2 \theta \quad (6.6)$$

It must be emphasized here that, because of calibration conditions, the result of the measurement should reflect the total power density near a source and not the electric or the magnetic power density only. Thus, contrary to formula (6.2), when formulas (6.5) and (6.6) were introduced, we assumed:

POWER DENSITY MEASUREMENT 117

$$S_r = S_E = S_H \qquad (6.7)$$

Comparing the power densities given by formulas (6.4), (6.5) and (6.6) we define the error of the power density measurement near the elemental E-source by the electric field measurement η_{EE} and by the magnetic field measurement η_{EH}:

and:
$$\eta_{EE} = \frac{|S_r - S_E|}{S_r} \qquad (6.8)$$

$$\eta_{EH} = \frac{|S_r - S_H|}{S_r} \qquad (6.9)$$

Formulas (6.8) and (6.9) are useful for the error of measurement estimations when the error value does not exceed, say, 15 percent. For larger values of the errors η_{EE} and η_{EH}, they become difficult to interpret and compare with other measurements. Meanwhile, the EMF strength and the power density measurements are some of the least accurate physical quantity measurements and often are accepted as satisfying the error value on the level of ±3 dB, or even ±6 dB. Figures 6.1 and 6.2 are the plotted errors δ_{EE} and δ_{EH} and are defined as:

$$\delta_{EE} = 10 \log \frac{|S_E|}{S_r} \qquad (6.10)$$

$$\delta_{EH} = 10 \log \frac{|S_H|}{S_r} \qquad (6.11)$$

118 ELECTROMAGNETIC FIELD MEASUREMENTS IN THE NEAR FIELD

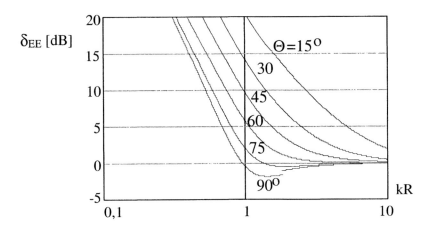

Figure 6.1. *Error δ_{EE} as a function of kR.*

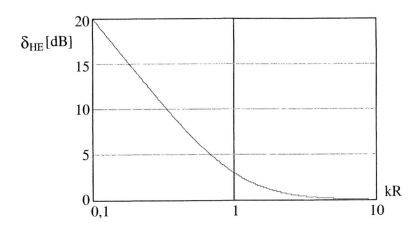

Figure 6.2. *Error δ_{EH} as a function of kR.*

Similar considerations for a source in the form of an elemental magnetic dipole, we have errors of the power density measurement in proximity to the source δ_{HE} and δ_{HH} as a result of the use (for the power density estimation around the dipole) of the **E** and **H** measurements respectively.

Instead of repeating calculations similar to those above, we will make the use of the analogy between E- and H-type sources (principle of the symmetry) that allows us to write:

$$\delta_{EE} = \delta_{HH}$$
$$\delta_{EH} = \delta_{HE} \qquad (6.12)$$

Our presented considerations may be summarized as follows:

1. The power density measurement in the near-field (especially in proximity to electrically small sources) using the electric or magnetic field measurement is burdened with an error whose value is dependent upon both the type and the structure of the source and the measured EMF component as well as the distance between the source and a point of observation.

2. For a source in the form of the elemental electric and magnetic dipole, the measurement error is always large for $\theta \to 0$ and it decreases for $\theta \to \pi/2$. While E-field is measured near the magnetic dipole and H-field near the electric dipole, the error is independent of θ. It is possible to assume that for $\theta \approx \pi/2$, the measurement errors do not exceed ± 6 dB for kR > 1.

3. Based on the curves shown in Figures 6.1 and 6.2, as well as assuming the smallest distances from a field source in which the measurements should be performed, it is possible to estimate the minimal frequency at which the measurement may be applied without the necessity of using additional correction factors (because of the deterministic character of the error, the factors may be analytically estimated for a known source type, measured EMF component, propagation geometry and distance).

4. It is possible to estimate the maximal error of the power density measurement using these methods in the neighborhood of the elemental sources, in accordance with the American Protection

Standard (measurement at a distance of 5 cm away from a source) or with Polish ones (30 cm off a source). In both cases, the error exceeds 6 dB when measurements are performed at 300 MHz. Of course, the error value increases as the frequency decreases (if the distance is kept constant).

The following two aspects of the considerations should be also emphasized:

1. The measurement method under consideration is widely used in power density meters available on the market; the meters 'assure' the power density measurements at frequencies even below 10 kHz without any explanation related to the interpretation of measurement results. We should not be surprised by the approach of the meters' manufacturers as the requirements are formulated in many standards (or standard proposals) that are prepared with participation of experts!

2. These considerations made it possible to estimate maximal method errors. However, it was shown [1] that measurements performed in proximity to physical sources (of finite dimensions) are not remarkably lower – the statement is, in some sense, evident in light of Maxwell's equations. The electric field is proportional to the charge of the source while the magnetic one is proportional to its derivative (current) when in close proximity to a source. This is a condition of the quasi-stationarity of the field. In the stationary field the E- and the H-field may be considered as independent of one another. An example of such situation is the EMF around overhead power transmission lines where the E-field appears while the line is connected to a voltage, but the H-field appears only if the line is loaded.

6.1.2. Power density measurement using the arithmetic mean of S_E and S_H measurement

Some protection standards define allowable levels of the sum of the electric and the magnetic power density. The procedure is equivalent to that in the introduction of this chapter, as the

POWER DENSITY MEASUREMENT 121

measured power density, the arithmetical mean of the electric and the magnetic power density. Let's try to estimate the increase in measurement accuracy it is possible to achieve in this way.

As shown above, the results of the power density measurement using the electric field measurement near a source with the magnetic field dominating and *vice versa* are underrated. At the same time, while the character of the source and the measured component are similar, the results are overstated. It might suggest that the measurement of the arithmetical mean of the values might lead to the improvement of the accuracy of the power density measurement.

As before, solving for the power density measurement error by the way of the arithmetical mean of S_E and S_H measurement we define in the form:

$$\eta_a = \frac{|2S_r - S_E - S_H|}{2S_r} \qquad (6.13)$$

Because of the reasons presented above, the formula (6.13) will be modified to the logarithmic form:

$$\delta_a = 10 \log \frac{|S_E + S_H|}{2S_r} \qquad (6.14)$$

Calculated values of the error δ_a are plotted in Figure 6.3.

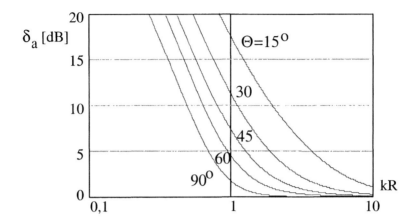

Figure 6.3. *Error δ_a as a function of kR.*

Because of the principle of symmetry, formula (6.14) and the results presented in Figure 6.3 are identical for the case of an elemental electric dipole and a magnetic one. If we compare the curves shown in Figures 6.1, 6.2 and 6.3, we can see that the error values in Figure 6.3 are somewhat smaller as compared to Figures 6.1 and 6.2. However, it is not a remarkable improvement and the construction of a device for such measurement would be complex and expensive.

Although it is not clear why both the power densities do not have 'equality of rights' [2], it is why the power density is defined as a sum of 1/6 S_H and 5/6 S_E. However, for this case an analysis of the accuracy improvement using a weighted arithmetic mean may be performed as well. The analysis was done and it shows that the measuring errors are very similar to those presented in the previous chapter.

6.1.3. Power density measurement as geometrical mean of the S_E and S_H measurement

Following the previously applied method, we will now estimate the power density measurement error, near a source of radiation, using the measurement of the geometrical mean of the S_E and S_H.

As in the case of the arithmetical mean measurement, the results are identical for both types of elemental sources.

The measurement error we define directly in the logarithmic form:

$$\delta_g = 10 \log \frac{\sqrt{|S_E S_H|}}{S_r} \quad (6.15)$$

The results of the error calculations are plotted in Figure 6.4. In comparison to the curves shown in Figure 6.3, a further leftward shift of the separate curves may be observed, which results in the further increase of the measurement accuracy. However, it is not a remarkable improvement, while the above-formulated conclusion, relating to the complexity of necessity for the measurement equipment and its cost, is still valid.

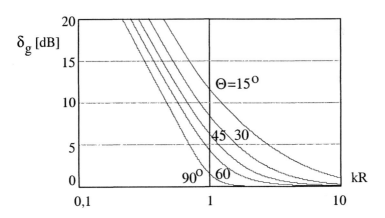

Figure 6.4. *Error δ_g as a function of kR.*

For elemental sources as well as for any open structure (with the exception of guided waves) is **E⊥H**. Based on formulas (6.1) and (6.2), this is written:

$$|S_m| = \sqrt{|S_E S_H|} \qquad (6.16)$$

Thus, the curves shown in Figure 6.4, apart from their role discussed above, show the ratio of the power density modulus to the power density radiated by a source. Formula (6.16) is true for any arbitrary EMF source. With no regard to the unknown physical sense of the imaginary part of the power density, it is possible to accept the geometrical mean measurement result as maximizing any possible value of the power density, however, a protection standard based on such a concept would be too restrictive.

6.2. Power density measurement using the antenna effect

The expression "antenna effect" is understood to mean the susceptibility of a loop antenna to the electric field. As mentioned in chapter 5.1 for near-field measurements, it is indispensable to use antennas that are much smaller than the minimal wavelength of the device's measuring band. They are sensitive only to **E** or **H**. It is true, however, only if, for instance, the diameter of a circular loop antenna $D = 2r_0$ fulfils the following condition:

$$D \leq 0.01 \lambda_{min} \qquad (6.17)$$

where λ_{min} = the shortest wavelength at which the antenna will work.

If formula (6.17) is not fulfilled, the emf induced in the antenna by **E** is not negligibly small as compared to the emf induced in the antenna by **H**. As a result, the antenna is useless for the measurements. Formula (6.17) is for a nonscreened and singly loaded loop and was introduced for the plane wave.

King [5.1] proposed a fascinating approach of the 'antenna effect' application in EMF measurements. The usefulness of the concept in different aspects of EMF measurements was analyzed in detail [3]. Its practical use was proven in a device for the near-field power density measurement. The essence of the concept is shown in Figure 6.5.

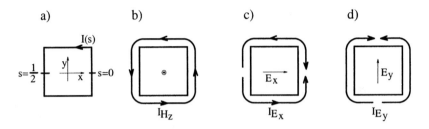

Figure 6.5. *Currents in a doubly loaded loop antenna.*

A quadrant loop antenna of the total circumference l, made of a conductor of diameter 2a, is located on the xy plane of the Cartesian coordinate system (Figure 6.5a). The resultant current induced in the loop is the sum of the magnetic field origin component, induced by magnetic field component H_z (Figure 6.5b) orthogonal to the plane of the antenna, and that of the electric field, induced by the electric field components E_x and E_y (Figure 6.5c and Figure 6.5d, respectively). If two symmetrically located loads are introduced to the circumference of the antenna, at $s = 0$ and $s = l/2$, the current at these points will be given by:

$$I(0) = I_{Hz} + I_{Ey} \tag{6.18}$$

$$I(l/2) = I_{Hz} - I_{Ey} \tag{6.19}$$

Formulas (6.17) and (6.18) are valid when the current component I_{Ex}, associated with E_x component, for $s = 0$ and for $s = l/2$ is equal to zero. This is a basic assumption of the method and the balance of the currents requires an appropriate antenna orientation in relation to the field components.

The principle of operation of a power density meter based on this concept is shown in Figure 6.6.

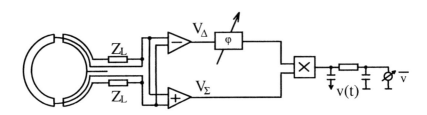

Figure 6.6. *Block diagram of a power density meter with a doubly loaded loop antenna.*

Output voltages from two loads, symmetrically immersed in the loop winding, lead to the inputs of two differential amplifiers. At their outputs we obtain the voltage of the sum of the input voltages V_Σ, as well as their difference V_Δ:

$$V_\Sigma = 2 K_E E_y \qquad (6.20)$$

$$V_\Delta = 2 K_H H_z \qquad (6.21)$$

where K_E and K_H = transmittance for the E and H field components.

These voltages are lead to a multiplier through a phase shift controller. The set is equipped with an output at which the voltage is proportional to the instantaneous value of the power density and another one at which, as a result of the integrating circuit use, the voltage is proportional to the mean value of the power density. This measurement concept is based directly upon the definition of the Poynting vector (formula 6.1).

The presented construction is an expansion of certain solutions that were applied for a long time in the radio direction finders working in the ADCOC system and similar ones. The aim of the work, carried out in this field at the Technical University of Wroclaw, was to prove the possibility of constructing a meter for the real part of the Poynting vector measurement using the doubly loaded loop antenna. It was also used to demonstrate that, although it is possible to use an arbitrary complicated combination of dipoles and single loaded loops, the use of a doubly (multiply) loaded loop allows remarkable limitation of the antenna effect. In the case of a doubly loaded loop, formula (6.17) may be modified to the form:

$$D \leq 0.15\lambda_{min} \qquad (6.17a)$$

Of course, the multiplication of the loads leads to the possibility of further increasing the electrical sizes of a loop antenna without necessarily taking into account the size limitations caused by the presence of the antenna effect. However, these solutions and designs are useful only in the case of far-field applications. This limitation is the result of the averaging of the measured magnetic field or power density at the surface of the antenna, which was discussed in Chapter 5.

Kanda applied the same concept in his construction of a power density meter [5]. Contrary to the design presented in Figure 6.6, he used a photonic link for the data transmission from both of the loop's loads to a measuring system. This solution allows remarkable improvement in the insulation of the antenna against the measured field and meter and multiplies the measured field interpretation possibilities due to preserved spectral and phase information.

This construction require several comments:

1. The use of a doubly-loaded loop antenna permits relatively simple power density measurement. At the same time it is possible to achieve very high sensitivity as a result of the permissible use of larger sized antennas (formula 6.17a) as well

as the introductory amplification of the sum and difference voltages.

2. The phase shifter applied in the meter as well as transmittances of both channels are a function of frequency (with regard to their phase and amplitude), and they considerably limit the applicability of the concept for discussed purposes, where widebandness of the measurement is one of its dominant advantages.

3. The transmission of signals from both loads to a meter may be troublesome because of deformation of the measured field by the leads. The use of a photonic link requires an increase of the probe's weight and/or more complex and expensive construction. However, photonic link technology will probably dominate in the future, as a result of photonic technology development advancements.

4. The design of the probe with a spherical directional pattern becomes more complex (see Chapter 7).

5. Formulas (6.17) and (6.17a) were introduced for the far-field (plane wave) conditions. In the near-field, the probe's size limitation is more rigorous and impossible to know beforehand for an arbitrary source. This is especially of concern for electrically small electric-type sources for which the following inequality is fulfilled:

$$|\mathbf{E}|/|\mathbf{H}| >> Z_0$$

6. The necessity of immersing the measuring antenna in the measured field in such a way that E_x is compensated for, is a factor making the measurement more difficult, especially measurements in the proximity of many non-correlated sources working at different frequencies as well as in the conditions of multipath propagation, not to mention about their temporal variations.

In order to illustrate problems relating to the structure of the presented meter and its work, we will consider the problem of frequency response matching of both the electric and magnetic part of the device. Formula (5.4) presents a mutual relation of sensitivities when no means of frequency response shaping was applied. This problem was discussed in Chapter 4, in relation to the electric field probes and in Chapter 5 to the magnetic ones. The formula implies the necessity of using these means to form a relationship of the sensitivities to both field components independent of the wavelength. The problem is much wider, however, it requires us to achieve appropriate transmittances for both EMF components; because of the calibration conditions, with the use of the TEM wave, their mutual relation should reflect the power relations in the plane wave. Apart from the necessity of having identical sensitivities for both field components it is important to have identical shapes of their frequency response as well.

In the most general case, the relation of transmittance within the measuring band of the electric field probe (creating a part of the power density meter) T_E to that of the magnetic field probe T_H may be defined in the form:

$$\frac{T_E}{T_H} = \frac{h_E \eta_E E}{h_H \eta_H H} \tag{6.22}$$

Where h_E, h_H = effective lengths of the electric and the magnetic antennas applied

η_E, η_H = attenuation factors of the detection circuitry of both the sensors

Because of the calibration's conditions, we must assume $E = Z_0 H = 120\pi H$.

Substituting h_E in accordance with formula (4.23) and h_H as given by formula (5.8) for h, $r_0 \ll \lambda$, we have:

$$\frac{T_E}{T_H} = \frac{\lambda h}{2\pi^2 r_0^2 n} \frac{\eta_E}{\eta_H} \qquad (6.23)$$

The maximal sizes of the antennas are limited to what was discussed in Chapters 4 and 5 for electric and magnetic antennas respectively. Based on acceptable sizes of these antennas, and resulting from them the permissible errors of both EMF components' measurement, we can write:

$$r_0 = a\,\lambda_{min}$$
$$h = ab\,\lambda_{min}$$

where a and b = constants, much smaller than unity
λ_{min} = the shortest wavelength of the measuring band

If we assume $\eta_E = 1$ and $\eta_H = 1$ for $\lambda = \lambda_{max}$ or $\eta_H = \lambda/\lambda_{max}$ (where λ_{max} = the longest wavelength of the measuring band), then after substitution of these assumptions in formula (6.23) we have:

$$\frac{T_E}{T_H} = \frac{b\lambda_{max}}{2\pi^2 a n \lambda_{min}} \qquad (6.24)$$

Because $T_E/T_H \equiv 1$, thus:

$$\frac{\lambda_{max}}{\lambda_{min}} = \frac{2\pi^2 a n}{b} \qquad (6.25)$$

Although the presented considerations are considerably simplified they allow us to formulate the following conclusions:

1. In order to obtain equivalent sensitivities for both field components it is possible to accept that the antennas applied should fulfill the condition: $h < r_0$,

2. The achievement of the wide range of measured frequencies is limited mainly by small effective length of the magnetic antenna and by relatively small sensitivity of the magnetic field measuring part of the device. It usually requires an artificial decrease of the sensitivity of its electrical part.

3. Because of the possibility of achieving a higher than necessary sensitivity in the probe and wider measuring band of the electric field, while sizes of the electric and the magnetic antennas are comparable, the construction of the magnetic (magnetic power density) part of the meter creates more problems. The sensitivity and the frequency response corrections are possible in both parts of the meter by selection of appropriate geometrical sizes and shapes of the antennas as well as with the use of RC bandpass filters.

4. As the discussion shows, the condition $h < r_0$ is fulfilled in the majority of cases. Thus, it is possible to assume that the error of the power density measurement with an E/H probe (error of the measurement not that of the method) is dominated by the error of the H-field measurement.

5. If $T_E \neq T_H$ in the function of frequency or in the function of measured value, the measuring error of the set increases. An example of the maximal error value is a case when the probe is entirely insensitive to the E- or H-field and, as a result, the power density is then established on the basis of only one component measurement.

The outlined issues are, perhaps, of excessively theoretical character. However, correct use of an arbitrary meter requires understanding of its work principles and, subsequently, limitations of its use and expected accuracy of measurement in different conditions.

6.3. Conclusions and comments

Widely applied methods of the electromagnetic power density measurement were presented and analyzed in this chapter. The considerations have a purely theoretical character and they were concerned with the field relations only. The source of the accuracy limitation of the measurement, by way of E- or H-field measurement, (error of the method) was demonstrated and the magnitude of the error for different combinations of source and measuring probe was estimated. Then a certain improvement of the accuracy, as a result of simultaneous measurement of E- and H- fields, and calculation of an arithmetical or geometrical mean was proposed. The measurements of the mean values may be realized by two, simultaneously independent measurements of these components using two different meters (one of them measures electric power density and the other magnetic one). Then the final result is obtained by computing the wanted mean or by designing a probe (meter) that would be simultaneously sensitive to both components, the calculations could be done by a processor in the meter and the ready result will be displayed. Construction of a meter that allows direct measurement of the real part of the Poynting vector was also outlined. Thus, there does exist the real possibility of measuring the power density in the near-field in any aspect and in almost arbitrary conditions. With no regard to technical efforts to design appropriate equipment, more and more precise meters available on the market and better understanding the problem a question arises:

What is the sense of the power density and its measurement?

The quantity "power density" has primarily been introduced to the technique, and it is still successfully exploited, in antennas and propagation considerations, especially at microwave frequencies. A typical example of the quantity use is in "radiolocation equation" or in the widely applied term "effective surface of an antenna." It seems that the latter has strongly influenced the impression of similarity between an antenna and the human body or other irradiated object. By simple multiplication of the power density by the effective surface of the absorber, the power (energy) absorbed by the object from EMF can be obtained. In the case of object irradiation by a plane wave, such an approach could be, at least, accepted. The errors of estimation, resulting from the measured field deformations caused by the object and the phenomenon of energy dividing into its absorbed and reflected parts, can be taken into account in the form of analytically estimated correction factors, dependent on the shape and electrical parameters of the object, polarization of the field (in relation to the object) and other factors. In near-field conditions such an approach, although theoretically possible, is nonsensical.

The analysis of presented curves, illustrating the accuracy of the currently used power density measurement methods, permits the statement that at frequencies below 300 MHz, the measurement is encumbered with remarkable measuring error (to remind: the error of the method). Because of the deterministic character of the error, it is fully possible to estimate appropriate correction factors in every situation of practical importance, introduce them into results of a measurement and in this way to achieve extended accuracy of the measurement. If, at least, to not take into account the possibility of using a meter that would make possible a direct measurement of the real part of the power density, because the cost of such a device, its complexity or because of unavailability of such devices on the market. As may be seen, if appropriate efforts (expenditures) are engaged it is possible to measure the quantity under almost any conditions. Apart from the aim of the presented considerations, which is the EMF metrology designated for surveying and monitoring of EM environment and resulting from the necessity of having at our disposal a means to perform quick and relatively simple measurements, we will return to the above formulated question in a bit wider aspect.

In order to answer the question we shall refer to the EMF properties of the simplest elements in electrical engineering, e.g., a capacitor and an inductor (we should focus our attention on the fact that the elemental electric and magnetic dipoles, that were assumed here as the basic sources in our considerations, are very similar to these elements in the electromagnetic aspect). The phase difference between the electric field and the magnetic one inside of these elements is equal to $\pi/2$ and, as a result, their vector product (expressed by formula 6.1) is equal to zero. It means that in the region en, electromagnetic energy radiation does not appear or that in the area losses of electromagnetic energy do not exist. The above said does not mean that energy transfer in the area is impossible at all. The best example here is a wide use of both elements for energy transfer as heating electrodes in such applications as dielectric or inductive heating as illustrated in Figure 6.7. However, the energy transfer requires a lossy medium inside the capacitor or the heating coil. We should add that the absorption model considered in Chapter 2 is such a situation while a lossy medium was immersed between the plates of a capacitor.

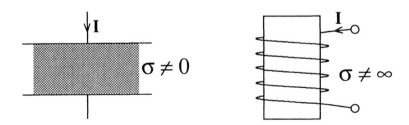

Figure 6.7. *Lossy medium allows dielectric (left) or induction (right) heating.*

Correct power density measurement, is without the measured field distortions (while it is one of the most important requirements here, and it will be briefly outlined in Chapter 8 but inside a capacitor or a solenoid must give zero-th result. By deduction we can widen the conclusion to the neighborhood of elemental (electrically small) sources and say that the measurement, in considered aspect, makes no sense because it does not reflect the phenomena related to the energy transfer. As completely reasonable as this is, however, the measurement of the temperature increase of a lossy dielectric (in the case of dielectric heating) or a lossy conductor or semiconductor (in the case of inductive heating), but the measurement is dependent upon many other influencing factors that were mentioned in Chapter 2 and they radically limit the usefulness of the method for discussed purposes.

The deliberations presented may be summarized as follows:
The power density measurement in the near-field (in the aspect considered), with no regard to the method applied, is senseless for frequencies below approximately 300 (1000) MHz. The same reservation should be applied to the power density, as a physical quantity, application in any protection standard and measuring device devoted to the measurement of frequencies below 300 MHz. Many of the protection standards and measuring devices have improved lately, but even newly proposed standards are not free of weaknesses. In order not to mislead the users, the physical quantity must not be applied in any aspect, even for illustration or comparison purposes. On the other hand, the use of the quantity on microwaves, at frequencies above 300 (1000) MHz is acceptable. However, even here it is necessary to take precautions. Remember the example presented in Chapter 2, where the far-field of a relatively (geometrically) small dish antenna is similar to that of very tall long-wave antenna.

Of course, every physical quantity can be applied in any, arbitrary way. The basic problem is *understanding* its denotation. That said, it does not change the above conclusion, on the contrary, it confirms that the limitations related to the use of this quantity must be rigorous. If the people involved in the preparation of protection standards, (i.e., the best experienced and most competent persons) have difficulties understanding these problems, then we can only guess the problems that would face ordinary inspectors of surveying-control services.

6.4. Bibliography

1. H. Trzaska, "Near-field Power Density Measurements," The First World Congress for Electricity and Magnetism in Biology and Medicine, Orlando, FL 1992. In: *Electricity and Magnetism in Biology and Medicine*, M. Blank (Editor), San Francisco Press, Inc., pp. 581–583.
2. P. Pirotte, "The Problematic ELF Since 1970 and the Actual Situation with CENELEC Pre-regulations," COST-244 WG Meeting, Athens 1995.
3. D. J. Bem, T. Wieckowski, "On the Measurement of Hazardous EM Field in Lossy Media Using a Small Loop Antenna," *Proc. 1981 Int'l. EMC Symp.*, Zurich, pp. 181–186.
4. D. Hoff, G. Monich, *Eine Sonde zur direkten Messung von Energiestrommungen im Nahfeld von Sende und Empfangsantennen* (in German). NTZ, No. 27/1974 Heft.8, pp. 313–318.
5. M. Kanda, "An Electromagnetic Near-Field Sensor for Simultaneous Electric and Magnetic Field Measurement," *IEEE Trans.*, Vol. EMC-26, No. 3/1974, pp. 102–110.

7

Directional Pattern Synthesis

When radio communication or propagation problems are discussed, an idealized understanding of the expression *linear* or *circular* (or more accurately, *elliptical*) polarization is usually used. At the same time, in order to simplify anyway complicated problems, it is 'forgotten' that even a linearly polarized wave propagating near the surface of a lossy medium changes to an elliptical polarization that is applied to measure the equivalent conductivity of the medium. In the case of multi-path propagation of a circularly (elliptically) polarized wave as a result of multi-path interference, a spatial rotation of the polarization plane may be observed. Thus, it may be called quasi-spherical or quasi-spherical polarization. Usually this occurs in the neighborhood of a complicated system of radiators (primary and secondary ones), when they are excited by an FM modulated signal or, as a result of phase difference changes of the rays converging on an observation point due to frequency changes or the Doppler effect.

Under the presented conditions, three spatial components of the measured EMF vectors may appear at a point of observation and have been taken into account when measuring errors due to spherical pattern changes of the electric field probe (Chapter 4) and that of the magnetic one (Chapter 5) were being estimated. This in no way changes the statement (Chapter 3.1) that at any arbitrarily selected moment of time and space there exists one and only one electric field vector and one magnetic field vector. The expression 'polarization' implies the possibility of drawing in the space, by the head of a vector, complex curves (elliptical polarization) or planes (ellipsoidal polarization). The complexity shows that metrological difficulties may occur that can lead to measurement interpretation problems with the results and, as a result, a decrease of the measurement accuracy.

The considerations presented in Chapter 4.4 lead to the following conclusions:

- An arbitrary radiating system that is small in relation to the wavelength and composed of elements sensitive to electric or magnetic fields has a sinusoidal directional pattern. A specific case of the 'antenna effect' use as an example of simultaneous sensitivity to both field components was presented in Chapter 6.2.
- It is impossible to construct an omni-directional antenna insensitive to EMF polarization using any number of arbitrarily oriented elements while they are connected to a common load
- In order to obtain a spherical (omni-directional) pattern it is necessary to use at least, three mutually orthogonal antennas and their output voltages should be summed after squaring them
- It is worth remembering that we discuss in detail the properties of omni-directional E-field probes. However, the designs of magnetic probes are almost identical, as are the directional patterns of electrically small electric and magnetic radiators (sensors).

7.1. A probe composed of linearly dependent elements

Let's consider a probe, designated only for elliptically polarized field measurements, composed of n identical dipoles intersecting in the center of the xy plane of a Cartesian coordinate system. The angles between them are identical and equal π/n. The first dipole crosses the axis y at angle γ as shown in Figure 7.1. The probe is illuminated by a monochromatic plane wave of angular frequency Ω propagating in z direction. Its electrical field components are given by:

$$E = 1_x A \sin \Omega t + 1_y B \cos \Omega t \qquad (7.1)$$

where A and B are amplitudes.

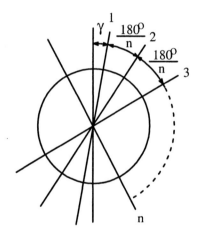

Figure 7.1. *System of n dipoles.*

If we assume that the sizes of the system are small in comparison to the wavelength, the influence of the mutual couplings between antennas is negligible, and all of the antennas are loaded with square-law detectors then, based on the above-formulated conclusions, it is possible to describe their directional patterns as sinusoidal. Thus, the square of the output voltage of the i-th antenna (v_i) is (7.2):

140 ELECTROMAGNETIC FIELD MEASUREMENTS IN THE NEAR FIELD

$$v_i^2 = \frac{h^2}{T} \int_0^T \left[A \sin \Omega t \cos (1_x, 1_i) + B \cos \Omega t \cos (1_y, 1_i) \right]^2 d(\Omega t) =$$

$$= \frac{h^2}{2} \left[A^2 \cos^2 (1_x, 1_i) + B^2 \sin^2 (1_x, 1_i) \right]$$

(7.2)

where 1_i = versor of i-th dipole.

In order to calculate the output voltage from n dipoles, we will use, especially for this purpose, the generalized Pythagorean theorem that can be formulated in the following form:

If, in a circle are inscribed n secants that intersect in the center of a circle, the angle between two adjoining secants is π/n and the first of them crosses the axis of symmetry of the circle at angle γ, then the sum of squares of the directional sines (and cosines) of the secants is constant and equals n/2. We write it in the form:

$$\sum_{i=1}^{n} \cos^2 \left(\frac{\pi}{n} i + \gamma \right) = n/2$$

$$\sum_{i=1}^{n} \sin^2 \left(\frac{\pi}{n} i + \gamma \right) = n/2$$

(7.3)

Formula (7.3) is valid for n ≥ 2. Denotations used in the formula are identical with those in Figure 7.1. We notice that when n = 2, the formulas are equivalent to the traditionally understood Pythagorean theorem.

If we take a sum of formula (7.2) making use of formula (7.3) for n detectors, we have the square of the output voltage V of the system. For elliptical polarization it is:

$$V^2 = \sum_{i=1}^{n} v_i^2 = \frac{nh^2}{4} \left(A^2 + B^2 \right)$$

(7.4)

For circular polarization A = B and:

$$V = Ah\sqrt{\frac{n}{2}} \qquad (7.5)$$

For linear polarization, B = 0 and:

$$V = Ah\frac{\sqrt{n}}{2} \qquad (7.6)$$

This consideration allows us to formulate the following two conclusions:

1. The results of formulas (7.5) and (7.6) show the difference of the probe output voltage depending upon the measured field polarization. However, regardless of the known voltage V, it is not possible to find *a priori* the magnitudes of A and B and then, the polarization of the field.

2. The output voltage is a function of n. Thus, when probes containing electric and magnetic antennas for the power density measurement are designed; it is indispensable to assume identical n for both EMF components.

In the presented estimations it was assumed that $h \ll \lambda$. It is possible to simplify the calculations. The simplified approach, however, leads to an error resulting from the radiation pattern changes when the ratio of h/λ increases. The problem was discussed in Chapter 4.4 and it may be assumed that the factor can be neglected with an accuracy of δ_{1E}.

The directional properties of the system were discussed for the probe illumination by the plane wave when the probe is designed for the near-field measurements. Changes of the spherical pattern as a function of R_0/h were discussed in detail in Chapter 4.4 and the considerations are valid in case of any combination of considered here antennas.

The equations in formula (7.3) are of purely trigonometric nature and their use for the pattern calculations is limited to values of n that are not very large and while mutual couplings between individual dipoles are negligibly small.

This discussion was presented to illustrate the problems but they are accurate enough for practical applications.

7.2. Spherical radiation pattern of an E/H probe

Examples of applied versions of E/H sensors are shown in Figure 7.2.

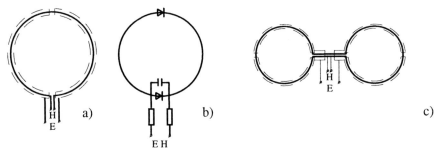

Figure 7.2. *Examples of E/H probes, a) set with divided electrostatic screen, b) probe with doubly loaded loop, and c) probe with two loop antennas.*

In the solutions shown in Figures 7.2a and c, the role of the electric field sensor and the magnetic one are separated and there appears the possibility of independent shaping of their frequency responses in accordance with the considerations presented in chapter 6.2. As far as the solution shown in Figure 7.3, and similar to those presented in Figure 6.6, these solutions allow full separation of functions of both sensors. The simplest solution shown in Figure 7.2b does not include such a possibility and, as a result, it may be used for the power density measurement only at a singular, selected frequency at which the calibration of the probe was performed. Its main advantage is the simplicity of its construction.

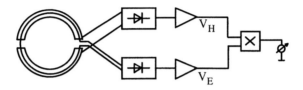

Figure 7.3. *Block diagram of the power density meter.*

All of the designs shown in Figure 7.2 are loaded with a common disadvantage – a very strong coupling between antennas of both EMF components and ineffective performance of the electrostatic screen divided into two parts (Figure 7.2a), while the solution with two loop antennas (Figure 7.2c) is too large, especially for near-field measurements where the effective length of the loops (even multiturn ones) are small in comparison to that of the electric dipole.

In Chapter 7.1 we analyzed the possibility of constructing an EMF sensor for elliptically polarized field measurement using a combination of linearly dependent antennas. This approach is applied to the available probes' solutions in order to increase their sensitivity (the output voltage is then proportional to the square root of the number of antennas applied — n). However, the practical solutions represent a slightly modified approach, i.e., the use of n quadrant antennas with one common arm. The common element is placed perpendicularly to the plane created by n unipoles. The detection diodes are connected between the common unipole and their respective unipoles at the plane.

144 ELECTROMAGNETIC FIELD MEASUREMENTS IN THE NEAR FIELD

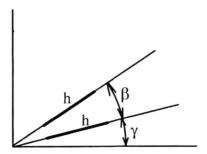

Figure 7.4. *The quadrant antenna.*

The electromotive force e at the output of a short quadrant antenna, as shown in Figure 7.4, under conditions of similar illumination as presented in Chapter 7.1, when illuminated from the direction of the bisector of the angle β, is:

$$e = C \cos\left(\gamma + \frac{\beta}{2}\right) \sin \frac{\beta}{2} \qquad (7.7)$$

where C = constant, other indications as in Figure 7.4.

If we assume in formula (7.7) that $\gamma = 0$ and $\beta = \pi/2$, which reflects the above-presented construction of the probe, we obtain a two-fold reduction of the emf induced in a quadrant antenna with relation to a symmetrical dipole of identical size. Thus, the solution's benefit is evident, especially if we take into account that the usual number of unipoles is n > 20. The directional pattern of an individual quadrant antenna is (in accordance with the above) sinusoidal and the direction of its maximum is identical to that of the bisecting line of the antenna angle β. Thus, pairs of the quadrant antennas, which create a common plane, are sensitive to the circular polarization in the plane. Thus, if n = 4, or a multiple of four, and any unipole is positioned symmetrically in relation to the others, it is possible to synthesize a spherical pattern of a quadrant system.

A similar concept has already been applied in the construction of an E/H sensor [1]. If we use only one loop antenna, immersed inside an electrostatic screen (e.g., as a half of the set shown in Figure

7.2c), and place three such loops on three separate planes of the Cartesian coordinate system in such a way that their inputs are located close to the center of the coordinate system, the system allows us to synthesize the spherical pattern for the magnetic field. This is simplified remarkably in relation to that containing three separate double loop probes as presented in Figure 7.2c and placed perpendicularly to each other. However, the use of electrostatic screens in these loops as arms of three quadrant antennas does not assure the spherical radiation pattern for the electric field. If we suppose that the shape of the radiation pattern for the magnetic field is satisfactorily uniform, then the screens of the loops can be replaced by unipoles of equivalent effective length h. To satisfy the requirement of a spheroid pattern for the electric field, we need to add to the system an auxiliary unipole that will be located symmetrically in relation to the previous ones, as shown in Figure 7.5.

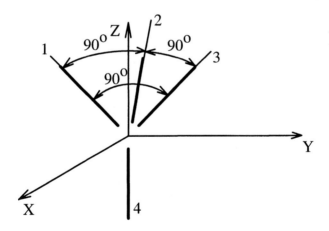

Figure 7.5. *Spatial orientation of unipoles in a modified triple quadrant antenna.*

The system of unipoles shown in Figure 7.5 contains six quadrant antennas that have common arms. Three of the quadrants are created successively by pairs of unipoles 1–3, while the next three quadrants are created by the unipole 4 and unipoles 1–3 respectively. The analysis shows that (if the length h of unipoles 1–3 are identical and they are infinitely thin) the length of unipole 4 (h_4) should be:

$$h_4 = h\left(2 - \sqrt{3}\right)$$

However, because of the thickness of unipoles 1–3 (in the role the electrostatic screens of the loop antennas), in order to obtain the spherical pattern of the electric field probe it is indispensable to enlarge unipole 4 by two and, depending upon the screens' sizes, we have:

$$h_4 = h\,(0{,}5 \div 0{,}75)$$

The outlined considerations are approximate and incomplete. They were added here to illustrate the methods and concepts of the spherical pattern synthesis. However, they were necessary to help us understand and evaluate the technical solutions available on the market, which (sometimes) look very attractive, but their construction and parameters may exclude their applicability in many measurements, especially those performed in the near-field.

7.3. A probe composed of three mutually perpendicular dipoles

In Chapter 4.4, we showed that in order to obtain an omnidirectional pattern it was necessary to use three mutually perpendicular antennas and sum their output voltages after squaring. Let's consider a system fulfilling these conditions. Three identical, mutually perpendicular dipoles a, b and c are located in the Cartesian coordinate system as shown in Figure 4.11. The system is illuminated by a monochromatic plane wave elliptically polarized. Its electric component may be expressed in the form [2]:

$$\mathbf{E} = \mathbf{1}_x A \sin \Omega t + \mathbf{1}_y B \cos \Omega t \cos \omega t + \mathbf{1}_z C \cos \Omega t \sin \omega t \quad (7.8)$$

where A, B and C = constants (amplitudes),
  ω = angular frequency of the polarization ellipse rotations, let $\Omega \gg \omega$,
  $\mathbf{1}_x, \mathbf{1}_y$ and $\mathbf{1}_z$ = versors of the coordinate system.

The emf e_i induced by the field in i-th dipole is expressed in the formula:

$$e_i = hA \sin \Omega t \cos(\mathbf{1}_x, \mathbf{1}_i) + hB \cos \Omega t \cos \omega t \cos(\mathbf{1}_y, \mathbf{1}_i) + \\ + hC \cos \Omega t \sin \omega t \cos(\mathbf{1}_z, \mathbf{1}_i) \quad (7.9)$$

where $\mathbf{1}_i$, i.e.: $\mathbf{1}_a, \mathbf{1}_b$ and $\mathbf{1}_c$ = versors of respective dipoles located in the Cartesian coordinate system 'a,' which is arbitrarily rotated in relation to the system shown in Figure 4.11. The procedure for squaring of the emf e_i may be performed twofold. We will discuss both possibilities.

7.3.1. The vector summation

For the vector summation we define the procedure based upon the summation of the output emf of the three dipoles e_i (i.e., e_a, e_b and e_c corresponding to the dipoles a, b and c respectively) after squaring them. The summation is realized using phase information conservation and without any additional flirtation procedures that would limit the measuring frequency band.

When quasi-ellipsoidal polarization is considered, the output voltage is equal to the sum of the three emf squared given by formula (7.9) and is expressed by complex formulas that are difficult to interpret. In order to simplify the considerations (and the final formulas), we assume that the measured field is polarized quasi-spherically, i.e., A = B = C. Thus, the sum of the emf squares, after some simple modifications, may be expressed as (7.10):

$$\begin{aligned}
V^2 = h^2 A^2 \Big\{ & \sin^2 \Omega t \big[\cos^2(1_x,1_a) + \cos^2(1_x,1_b) + \cos^2(1_x,1_c)\big] + \\
& + \cos^2 \Omega t \cos^2 \omega t \big[\cos^2(1_y,1_a) + \cos^2(1_y,1_b) + \cos^2(1_y,1_c)\big] + \\
& + \cos^2 \Omega t \sin^2 \omega t \big[\cos^2(1_z,1_a) + \cos^2(1_z,1_b) + \cos^2(1_z,1_c)\big] + \\
& + 2\sin \Omega t \cos \Omega t \cos \omega t \big[\cos(1_x,1_a)\cos(1_y,1_a) + \\
& \quad + \cos(1_x,1_b)\cos(1_y,1_b) + \cos(1_x,1_c)\cos(1_y,1_c)\big] + \\
& + 2\sin \Omega t \cos \Omega t \sin \omega t \big[\cos(1_x,1_a)\cos(1_z,1_a) + \\
& \quad + \cos(1_x,1_b)\cos(1_z,1_b) + \cos(1_x,1_c)\cos(1_z,1_c)\big] + \\
& + 2\cos^2 \Omega t \sin \omega t \cos \omega t \big[\cos(1_y,1_a)\cos(1_z,1_a) + \\
& \quad + \cos(1_y,1_b)\cos(1_z,1_b) + \cos(1_y,1_c)\cos(1_z,1_c)\big]\Big\}
\end{aligned}$$

Note that the expressions in the first three brackets represent sums of the squares of directional cosines of two rectangular coordinate systems rotated in relation one another and the sums are equal to unity. The expressions in the next three brackets represent products of respective terms of two rows or two columns of matrices of directional cosines of one coordinate system in relation to the other and they are by definition equal to zero. After some transformations we have:

$$V = h A \qquad (7.11)$$

For a circular polarization there is no rotation of the polarization plane ($\omega = 0$) and after some transformations, similar to the above and reduced to the plane of polarization, we have a result identical to that given by formula (7.11). For the linear polarization there is only one spatial component of the E-field and E is given by formula (7.8) in which, for example, B = C = 0. Taking this into account for the above considerations we have for the linear polarization:

DIRECTIONAL PATTERN SYNTHESIS

$$V = h A \sin \Omega t \qquad (7.11a)$$

Independent of the detector type used in the sensing system for the linear polarization we have a result identical to that obtained for quasi-spherical and circular polarization using peak value detection and it differs from it by 3 dB when RMS detection (or rather detection of the mean value of V^2) is applied.

7.3.2. A DC summation

For the direct current summation we will define the procedure based upon respective dipoles loaded with the RMS detectors and a summation of their output voltages after filtration of the alternating components. Presently, this procedure is widely applied in omni-directional EMF probe construction, designated for measurements relating to labor safety and the protection of the natural EM environment as well as within the entire EMC area.

We will calculate the square of the output voltage of the i-th detector using the value of emf e_i given by formula (7.9):

$$\begin{aligned} V_i^2 &= \frac{h^2}{T\tau} \int_0^T \int_0^\tau e_i^2 \, d(\Omega t) \, d(\omega t) \\ &= \frac{h^2}{4} \left[A^2 \cos^2(\mathbf{1}_x, \mathbf{1}_i) + \frac{B^2}{2} \cos^2(\mathbf{1}_y, \mathbf{1}_i) + \frac{C^2}{2} \cos^2(\mathbf{1}_z, \mathbf{1}_i) \right] \end{aligned} \qquad (7.12)$$

where T and τ = periods equivalent to angular frequencies Ω and ω respectively.

After summation of output voltages V_i of three dipoles we have:

For quasi-ellipsoidal polarization:

$$V = \frac{h}{2} \sqrt{A^2 + \frac{B^2 + C^2}{2}} \qquad (7.13)$$

For the quasi-spherical polarization:

$$V = \frac{hA}{\sqrt{2}} \qquad (7.14)$$

For the elliptical polarization:

$$V = \frac{h}{2}\sqrt{A^2 + B^2} \qquad (7.15)$$

And for the linear polarization:

$$V = \frac{hA}{2} \qquad (7.16)$$

We should note here that similar results lead to a procedure of RMS detection of the output voltages when vector summation is used (formulas 7.11 and 7.11a). The difference between the summations (in a constant, nonessential from the presented considerations point of view) is a result of the assumption of the half-period detection.

7.3.3. Achromatic fields measurement

Until now we have considered directional patterns of several probe systems illuminated with a harmonic field. Very often it is necessary to perform EMF measurements in conditions where M spectral components appear simultaneously. The m-th electric field component we define in the folowing form (7.17):

$$E_m = 1_{xm} A_m \sin \Omega_m t + 1_{ym} B_m \cos \Omega_m t \cos \omega_m t + 1_{zm} C_m \cos \Omega_m t \sin \omega_m t$$

Formula (7.17) is identical in its shape to formula (7.8) and the only exception is the index 'm,' which differentiates the m-th spectral component. The proposed notation does not emphasize direction, amplitude or frequencies of the M spectral fringes of the E-field that appear in a specific point in space (the point where an EMF probe is placed — the probe is composed of three mutually perpendicular dipoles). The other assumptions here are identical with those accepted before. However, in order to obtain the required results of estimations an additional assumption here is indispensable, i.e., the assumption of the flat frequency response of the probe. The latter means that the effective length h is a frequency independent, constant value. The requirement is easy to fulfil when the measured spectrum appears only at the spectral fringes in the probe's medium frequency band (that was discussed in Chapters 4 and 5). When the condition is not fulfilled in formula (7.17), it is necessary to introduce multipliers that would normalize the transmittance of the probe to that of its medium frequency range.

The resultant field, illuminating the measuring device, is equal to the sum of its M components (frequency fringes). These components are of arbitrary frequencies, without the necessity of any harmonic relation between them that could suggest phase synchronization or a similar phenomena. As noted previously, the spatial location of any fringe may be arbitrary and should not affect the results of the calculations (measurements). Then the calculation procedure is identical to that applied in Chapter 7.3.2.

The electromotive force e_i induced by EMF in the i-th dipole is (7.18):

$$e_i = h \left\{ \sum_{m=1}^{M} \left[A_m \sin\Omega_m t \cos(\mathbf{1}_{xm}, \mathbf{1}_i) + B_m \cos\Omega_m t \cos\omega_m t \cos(\mathbf{1}_{ym}, \mathbf{1}_i) + C_m \cos\Omega_m t \sin\omega_m t \cos(\mathbf{1}_{zm}, \mathbf{1}_i) \right] \right\}$$

Making use of this formula we calculate the square of the output voltage of i-th detector:

$$V_i^2 = \frac{h^2}{T_m \tau_m} \int_0^{T_m} \int_0^{\tau_m} e_i^2 \, d(\Omega_m t) \, d(\omega_m t) \tag{7.19}$$

Exchanging the order of the summation and the integration we write (7.20):

$$V_i^2 = \frac{h^2}{4} \sum_{m=1}^{M} \left[A_m^2 \cos^2(\mathbf{1}_{xm}, \mathbf{1}_i) + \frac{B_m^2}{2} \cos^2(\mathbf{1}_{ym}, \mathbf{1}_i) + \frac{C_m^2}{2} \cos^2(\mathbf{1}_{zm}, \mathbf{1}_i) \right]$$

Now we reduce the formula (7.20) for different polarizations:

When the M signals are polarized quasi-ellipsoidally the probe output voltage is:

$$V = \frac{h}{2} \sqrt{\sum_{m=1}^{M} \left[A_m^2 + \frac{B_m^2 + C_m^2}{2} \right]} \tag{7.21}$$

For M signals polarized quasi-spherically:

$$V = \frac{h}{\sqrt{2}} \sqrt{\sum_{m=1}^{M} A_m^2} \tag{7.22}$$

For M signals polarized elliptically:

$$V = \frac{h}{2} \sqrt{\sum_{m=1}^{M} \left(A_m^2 + B_m^2 \right)} \tag{7.23}$$

And for M signals polarized linearly:

$$V = \frac{h}{2} \sqrt{\sum_{m=1}^{M} A_m^2} \tag{7.24}$$

This case points out that the results obtained using vector summation and when RMS detection is applied are similar, which confirms once more the equivalence of both procedures.

A more important conclusion is that the voltages given by formulas (7.21) and (7.22) are the measure of the effective (RMS) strength of the field containing more than one spectral fringe. Thus, it is not a sum of the effective magnitudes but the effective magnitude of the sum (resultant field).

7.4. Comments and conclusions

Detailed considerations of the polarization problems result from the need to understand and apply the omni-directional probes as well as their design and construction. These needs are the result of:

- The specificity of the polarization intricacies, especially in the near-field
- The necessity of understanding the polarization phenomena in order to select optimal procedures when measurements are planned and for interpretation of their results
- The development of the ability to select an appropriate probe and meter for specific measurement conditions
- The accomplishment of a skill crucial to evaluation of the market's offers

The directional properties of a probe have a fundamental importance for it possible use for variously polarized EMF measurements. We will briefly summarize below the application ranges as well as advantages and disadvantages of probes with selected directional properties.

7.4.1. Linear EMF polarization, sinusoidal pattern of the probe

This combination is especially characteristic for EMF measurements in the far-field. In the near-field who knows if the polarization is linear? Here is an advantage of the probe with a sinusoidal pattern: it makes it possible to check whether the EMF polarization is linear and, if it is, permits localization of the spatial orientation of the E (or H) vector. Directional properties of the probe allow us to localize an unknown source of radiation. In the case of linear polarization, the maximal indications of a meter appear when the measuring dipole is placed parallel to the E vector (formula 4.37). On the other hand, when the dipole is placed in a plane perpendicular to the E vector, the indications should be minimal, the latter is valid for an arbitrary position (rotation) of the dipole on the plane.

7.4.2. Elliptical EMF polarization, sinusoidal pattern of the probe

A good example of an EMF stable in time and space elliptical polarization, is the field under high voltage transmission lines. The polarization plane here is perpendicular to the line. The measurement using a sinusoidal probe allows us to find parameters of the polarization ellipse and, thus, the maximal value of E (and H) field strength. However, the measurement of the effective (RMS) value of the field requires two measurements and then calculation of the result of the measurement as a square root of the sum of squares of the two results. Both measurements should be performed on the polarization plane in two, mutually perpendicular positions of the probe. If there is doubt regarding the field polarization, it is easy enough to check. The indications of the meter vanish in the direction perpendicular to the polarization plane, i.e., parallel to the wires of the line.

7.4.3. Ellipsoidal (spherical) EMF polarization, sinusoidal probe

The denotation 'ellipsoidal' (spherical) polarization, as it has already been mentioned, has no physical sense. However, it is a convenient description of a situation where the polarization ellipse rotates in a space with a frequency (that may be a function of time and space), as discussed in detail in Chapter 7.3. Investigated EMF here has three, non-zero, spatial components. It leads to non-zero indications of a meter while the measuring probe is arbitrarily oriented in space. The advantage of the sinusoidal probe use is here, again, the possibility of measuring the spatial properties of the investigated field and finding the maximal strength (as to magnitude and direction) of the E (and H) vector. But the determination of the effective value (RMS) involves the same procedure as above with its expansion to the third dimension.

7.4.4. Linear or elliptical polarization, circular pattern of the probe

The probe with a circular pattern still has not been mentioned. The simplest example of the probe consists of two mutually perpendicular dipoles loaded with square-law detectors. During measurements of linearly or elliptically polarized fields using of the probe, it is enough to immerse the probe in the field and to orient it in such a way to as to notice maximal indications of the meter. It is certainly possible to use this probe for polarization evaluation, i.e., to prove if the polarization is linear or elliptical (minimal meter's indications, while the E vector of the investigated field is perpendicular to the plane where the measuring antenna of the probe is located, suggest linear polarization). However, it is impossible to absolutely determine whether the polarization is elliptical or quasi-ellipsoidal and there does not exist a simple procedure that would allow the use of the probe in conditions of polarization unknown *a priori*. This feature of the probe eliminates its practical applicability for our purposes with one exception: measurements in the proximity of overhead high voltage transmission lines where the polarization is well known.

7.4.5. Arbitrary polarization, spherical probe

This measurement consists only of introducing a probe into the measured field and reading the indications of the meter. For surveying or monitoring services this procedure has the same advantages as the probe. It is necessary to remember, however, that the probe does not permit us to find any information on the polarization properties of the investigated field, to measure maximal intensity of a spatial field component, or to detect a direction to an unknown field source.

7.4.6. Unknown probe

Given the number of manufacturers competing in the market and the battle to acquire customers by offering the lowest prices, not everything presented on offered meters in their manuals or announcements is clear or, even, true. It may be useful to prove the directional properties of the probes ourselves. The simplest way is to use a linearly polarized field generated between two plates (E-field) or the simplest circular coil (H-field). Using the procedures presented above, it is possible to prove the directional properties of a probe. A sinusoidal probe should have only one maximum, a probe with a circular pattern should show no indication of changes when it is rotated in such way that E-vector lies in the polarization plane of the probe, and the omni-directional probe should be insensitive (within limits indicated in the manual) to its arbitrary rotations in the field. The test should be performed at low frequency if possible.

As we have shown, an optimal design of the measuring probe does not exist. Although there have been available on the market meters that allow to measurement one, two or three spatial components of the field that is equivalent to all of the probes discussed above. However, because of their complexity and the possibility of making a mistake during measurements as well as their prices, they have not found wide use until now.

Not taking into consideration or ignorance of the polarization problems and inappropriate choice of a measuring probe (in this aspect) may lead to measurement errors exceeding 50% that, even with taking into account low 'resolution' of bioeffects and generally low accuracy of the field measurements, may be unacceptable. We should remember that all the considerations, regarding the directional properties, were taken into account for plane wave illumination conditions. The above-mentioned measurement error (50%) was also estimated for the conditions when the measuring probe separated enough that only polarization could dominate for its estimation. When measurements are performed in proximity to primary or secondary radiation sources, the error may be much bigger due to mutual coupling of the source and probe (that was discussed in Chapters 4 and 5) and other factors as well.

7.5. Bibliography

1. T. Babij, H. Trzaska, "Power Density Measurements in the Near-field," IMPE 1976, Microwave Power Symp., Leuven, Belgium. Publ.: *Journ.of Microwave Power*, No. 2/1976, pp. 197-198.

2. E. Grudzinski, H. Trzaska, "Polarization Problems in Near-field EMF Measurements," *36 Internationales Wissenschaftlisches Kolloquium*, Ilmenau 1991.

8

Other Factors Limiting Measurement Accuracy

In previous chapters, we discussed in detail the most typical and frequent factors limiting EMF measurement accuracy. They resulted mainly from the type of antenna applied in an EMF probe, its size, and mutual couplings between the antenna and a radiation source. In this chapter we will briefly discuss the influence of thermal drift upon the probe parameters, the role of its dynamic characteristics, and the deformation of the measured field by a person performing the measurements, the meter, resonant phenomena, and others.

8.1. Thermal stability of a field meter

The meter parameters are a function of temperature as well because of the strong thermal dependence of semiconductor device parameters applied in EMF probes mainly for the measured signal detection (diode, magneto-diode, Hall-cell, thermocouple). The thermal instability may be a source of remarkable measurement error. The errors (as the deterministic ones) may be limited to a considerable degree by the use of appropriate correction factors (calculated or measured) or by the choice of measurement conditions that would assure error minimization. Although measuring device manufacturers do their best to obtain the most acceptable parameters (including thermal parameters) for their tool, more and more extremely simplified devices are appearing on the market. These devices can assure a comparatively low price but sometimes it is at the expense of quality. In order to enable a user to evaluate these devices under the influence of extreme conditions, the role of quality (or lack thereof) must be examined.

The subject of our analysis will be a probe for an electric field measurement containing a dipole antenna loaded with a diode detector. This is the simplest version of the probe and the most susceptible to the influence of temperature changes. However, similar probes are widely used for electric field and power density measurement. The design of the magnetic field probe in many ways parallels that of the electric field probe and both are susceptible to ambient temperature variations. Although any element applied in an arbitrary type measuring probe is more or less susceptible to temperature changes, the detection diode is the most sensitive one. In order to improve thermal stability, a thermocouple is sometimes used instead of the diode; the diode is introductory polarized, and bridge diode detectors and other approaches are applied. Its important advantage is in the identical shape of the dynamic characteristics through a full measuring range that ensures true RMS indications. However, the range is very limited, its sensitivity is insufficient, and it can be easily damaged when overloaded. As mentioned above, we would like to illustrate the maximal values of the (and not just thermal) errors and explain why the simplest solution was accepted for the analysis.

In Chapter 4, we discussed how the parameters of the detector (diode) may affect the transmittance within the measuring frequency range (formula 4.14), and as a result the sensitivity of the meter, the shape of the transmittance in lower frequency range (formula 4.13), and the lower corner frequency (formulas 4.15 and 4.15a). The importance of these effects will be outlined consecutively.

8.1.1. Sensitivity alternations within the measuring band

The alternation of the meter sensitivity due to the temperature changes we define as follows:

$$\delta_1 = \frac{T_0 - T'_0}{T_0} \quad (8.1)$$

where

T_0 = transmittance of the probe within the measuring band in the temperature of the probe calibration

T'_0 = transmittance of the probe in the ambient temperature where the measurements are performed.

In light of formula (4.14), the transmittance T_0, is a function of the detector input capacitance. The capacitance consists of the parasitic capacitance of the montage, capacitance of the filter and the diode capacitances: the diffusion capacitance C_D, and the junction capacitance C_J. For instance, the capacitance C_D is given approximately by [1]:

$$C_D = \frac{q}{kT}(I_0 + I_s)\frac{\tau}{2} = g_d \frac{\tau}{2} \quad (8.2)$$

where q = the electron charge
k = Boltzmann constant
T = temperature
τ = carrier's relaxation time
I_0 = DC component of the diode current
I_s = saturation current
g_d = dynamic conductance of the diode

while

$$I_s \approx I_{s0} \exp(b \, \Delta T) \quad (8.3)$$

I_{s0} = saturation current at the starting temperature
ΔT = temperature rise
b = constant (b ≈ 0.07/K for germanium diodes and 0.1/K for silicon ones)

The subject of the temperature influence is also the capacitance C_J and the dynamic characteristics of the diode. While the dependence of the diode's specific parameters as a function of temperature is complex, its parameters also depend on the selected working point of the diode and its loading resistance. Usually in diode detectors and especially in diode rectifiers, it is assumed that $I_s \ll I_0$, which allows the simplification of formula (8.2) and indicates a decrease of the temperature effects on the diode parameters. In our case (which results from the diode loading with a transparent, high resistivity transmission line) the simplification is not acceptable, especially in the most sensitive measuring ranges. As a result, problems appear with stabilization of the detector's thermal conditions and there is an alternation of the probe parameters when the temperature changes.

Returning to the analysis of meter sensitivity and thermal stability, formula (4.14) correctly defines transmittance for a "high frequency" case. Here, the temperature affects capacitances C_D and C_J, or more precisely, the change of these capacitances in relation to the probe's other capacitances. By artificial extension of the detector's capacitance, it is possible to improve the thermal stability, which will be followed by a lowering of the lower corner frequency. However, it will happen at the expense of the sensitivity reduction. Here, however, there is a DC aspect of the problem as well. The DC part of the probe is not free of thermal instabilities. Based on Thevenin's theorem, it is apparent that the DC part of the probe has temperature instabilities for the following reasons. First, the probe is a low-current device due to its high resistive load. Second, we should consider the detector as the DC source of conductance given by formula (8.2), and the third takes into account formula (8.3).

A systematic approach to the problem is not presented here because of its complex character and dependence on many factors

OTHER FACTORS LIMITING MEASUREMENT ACCURACY 163

including selected probe configuration and the protective means applied. Detailed discussions of this approach are available in the literature [2, 3, and 4]. In order to summarize our discussion and to illustrate the scale of the problem, Figure 9.1 shows correction curves applied in the EMF meter type NFM-1 manufactured in the former German Democratic Republic and still applied in many Eastern European countries [5] (the problem is examined in various publications, e.g., [6]). Respective curves, prepared experimentally for different magnitudes of measured field, give the correction factors that are to be multiplied by the value indicated on the meter to obtain the measurement result. We should notice the increase of the meter's sensitivity with the temperature rise. The factors' maximal magnitudes are as much as 2%/K while the field intensity factor is about 5 V/m. Note that the influence of the temperature upon the sensitivity vanishes for larger intensities of measured field and the transition of the detector's working point to the linear part of its dynamic characteristics. Contrary to the applied practice, the curves should be periodically corrected because of variations induced by the aging process; moreover, the curves were prepared *en masse* for the meters while the thermal stability is a function not only of the type of the detection diode but also its specimen and its "history."

164 ELECTROMAGNETIC FIELD MEASUREMENTS IN THE NEAR FIELD

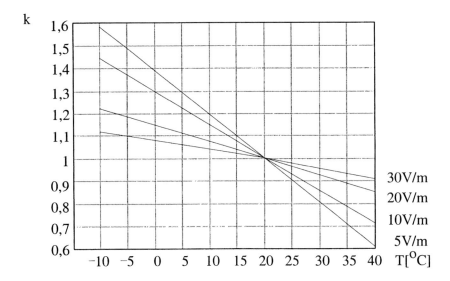

Figure 8.1. *Correction coefficients of the NFM-1 type field meter.*

8.1.2. Shape changes of the frequency response and the lower corner frequency

With the aim of attaining temperature dependence of the probe sensitivity not only as a function of the detector capacitance but also as a function of its conductance, it was necessary to quote formulas (8.2) and (8.3). The dependence of the lower corner frequency of both these parameters directly results from formulas defining the frequency (4.15 and 4.15a). Apart from the above comments related to the components of the detector capacitance, we should notice that in these formulas, R_d represents parallel connection of the dynamic resistance of the diode and the loading resistance. In other words:

$$\frac{1}{R_d} = g_d + \frac{1}{R_t + R_i} \qquad (8.4)$$

where R_t = resistance of the transparent line, and
R_i = input resistance of the line load (a DC voltmeter)

The resistance, as shown in formula (4.13), is of fundamental importance for the shape of the frequency response in the low frequency range.

In general, changes of the frequency response outside the measuring band are as important as those within the band both for the accuracy of the measurement and the sensitivity (transmittance) within the band. Let's summarize our thoughts:

- In a thermally non-compensated probe (considered here), the thermal changes of the lower corner frequency and the shape of the frequency response (sensitivity) within the low frequency range are the same order of magnitude as the sensitivity changes shown above.
- However, the thermal compensation of the sensitivity within the measuring band is not enough to obtain the compensation outside the band.
- From this we can draw the conclusion that there is a hidden a source of possible measuring faults when a probe is used (making use of the known shape of its frequency response outside the measuring band) for measurements outside the band. Meter manuals sometimes suggest (for instance, in the NFM-1 meter) in that this occurs in relation to frequencies below the band.
- Experience shows that all the thermal effects depend on the type of the diode applied and, may even, be different between sample one and sample two of the same type. Aging effects and various external factors can have a remarkable influence. For instance, overheating the diode during probe assembly or as a result of overloading affects both the dynamic characteristics of the detector and the thermal parameters of the diode; after overheating, the diode parameters drift over a relatively long period of time and they never return to their initial state.
- It is important to notice that both the shape of the frequency response and the lower corner frequency are functions of the

measured field strength magnitude; the dependence is especially noticeable if the changes of the lower corner frequency are compared to the field rate δ_{fl} as defined by:

$$\delta_{fl} = \frac{\Delta f_1}{f_1} \frac{1}{\Delta E} \tag{8.5}$$

where Δf_1 and ΔE = changes of the lower corner frequency and the rate of the measured E-field, respectively.

The measured magnitudes of δ_{fl} are [7] 0.01 – 1% m/V, for the detector with a germanium diode and 0.05 – 2% m/V for a detector with a Schottky barrier diode.

The presented effects are a function of the measured value; they are the strongest in the square-law part of the characteristics and vanish after the detector transitions to the linear part of the characteristics. The phenomenon is illustrated in Figure 8.2 [8].

OTHER FACTORS LIMITING MEASUREMENT ACCURACY 167

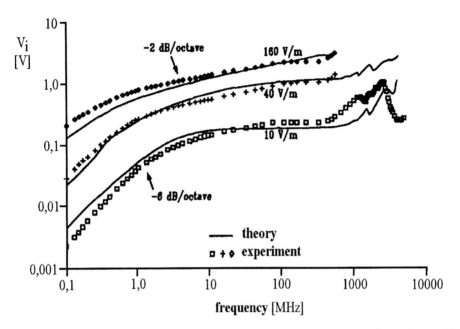

Figure 8.2. *Frequency response alternations as a function of measured E-field [8].*

8.2. The dynamic characteristics of the detector

The static characteristics of an ideal semiconducting diode in the range of permissible currents (i) and voltages (V_d) is given by:

$$i = I_s \left[\exp\left(\frac{q}{kT} V_d\right) - 1 \right] \tag{8.6}$$

The analytical calculation based on formula (8.6) of the DC voltage's magnitude at the output of the probe, i.e., the voltage at the input of a DC voltmeter represented by its input resistance R_i (Figure 4.3), requires the solution of a non-linear differential equation of the first order. We will cite here results of considerations presented by M. Kanda, performed for a monochromatic exciting signal and stabilized thermal conditions, based on the solution illustrated by P. F. Wacker [9]. We will complete the solution with terms for use of a frequency response shaping filter and division of the detector output voltage on the divider created by the transparent line resistance and the voltmeter's input resistance.

For small values of V_d, the DC voltage at the voltmeter input V_i is:

$$V_i = -\frac{q}{4kT}\left(\frac{e_A C_A}{C_A + C_d + C_f}\right)^2 \frac{R_i}{R_i + R_t} \quad (8.7)$$

Whereas, for large values of the voltage V_d:

$$V_i = -\frac{e_A C_A}{C_A + C_d + C_f} \frac{R_i}{R_i + R_t} \quad (8.8)$$

(Indications in the formulas as shown in Figure 4.3).

Based on these formulas, the probe's dynamic characteristics consist of three segments. These are: low values of V_d (for which it exhibits a square-law characteristic, as given by formula 8.7), high voltage V_d (where the characteristic is linear), and medium voltage V_d (where the shape of the characteristic is transitional from square-law to linear). The measured characteristic $V_i = f(E)$ for a typical specimen of an AE-1 probe is shown in Figure 8.3.

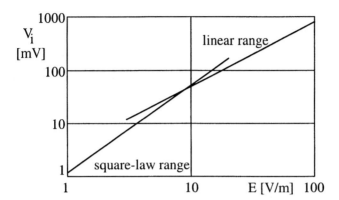

Figure 8.3. *Dynamic characteristics of an AE-3 probe measured for $T = 21°C$.*

The probe sensitivity alternations due to thermal changes in its dynamic characteristics were outlined in chapter 8.1.1. In order to emphasize the importance of temperature variations, formula (8.6) was quoted. If we omit the relatively narrow transitional segment, Figure 8.3 shows the change from square-law to linear for the probe's (detector) dynamic characteristics that are separate from the probe's susceptibility to temperature variations. The characteristic shape is completely non-essential when monochromatic harmonic fields are measured. But the phenomenon is important when measurements are performed in complex achromatic fields and in pulsed fields. We will illustrate this with several simple examples.

8.2.1. The AM modulated field with square-law detection

The amplitude modulated voltage V_m is shown in the form:

$$V_m = A(1 + m\cos\omega t)\cos\Omega t \qquad (8.9)$$

where A = amplitude
 m = modulation depth
 ω and Ω = angular frequencies of the modulating signal and the carrier wave, respectively

If the signal is detected with a square-law detector, its effective voltage V_{ieff} is given by:

$$V_{ieff} = \frac{A}{\sqrt{2}} \sqrt{1 + \frac{m^2}{2}}$$

The first term of the formula represents the effective value of the harmonic carrier wave while the rooted terms show the role of the modulation depth. The term varies from unity (for m = 0) to about 1.25 (for full modulation, i.e., for m = 1).

8.2.2. The AM modulated field with peak value detection

If we assume that in formula (8.9) the magnitudes of both cosines equal unity, we obtain the DC voltage amplitude at the output of the linear detector V_{ia}:

$$V_{ia} = A(1 + m)$$

If we assume m = 1, the amplitude of the voltage equals 2A.

8.2.3. Detection of a manipulated CW signal

Let's repeat the above considerations for a periodic signal of time duration T in which the carrier wave appears during period t. Consequently, we have for the square-law detection:

$$V_{ieff} = A\sqrt{\frac{\tau}{T}}$$

and for the linear detection:

$$V_{ia} = A$$

Each of these estimations provides correct results for a meter with a diode detector-equipped probe. For our purposes, both the effective and the peak values are of concern. The problem is in the necessity of precisely reducing the probe's measuring range to limits within which the dynamic characteristics are known. For economical reasons, the problem is often neglected. As a result, very large measuring errors may appear when the measurements of non-monochromatic fields and especially pulsed fields, are performed. If, in the case of the AM modulation we assume that a two fold difference between RMS and peak value measurement is acceptable, for the case of pulsed fields, the measurement error may reach an arbitrary value. Precise demarcation of the linear and the square-law segments of the dynamic characteristics may successfully limit these errors. This procedure can be done for any type of meter by analyzing the properties of the meter's indicator scale or its calibration curves (if applicable).

8.3. Measured field deformations

When a measurement is performed, the measuring probe must be immersed into the measured field. The probe is usually accompanied by the meter (indicating device) and a measuring person. Any material placed in a homogeneous field causes field deformation. Depending upon the electrical size and properties of the medium, the deformation may mean a change in the spatial distribution of the equipotential lines, reflection, or diffraction. The latter two phenomena usually appear when the medium's sizes are equal to or larger in comparison to the wavelength while the former is specific to (quasi-) static fields [10]. The problem must be known and understood by the people performing measurements and it must be mentioned in methodology textbooks and in appropriate standards for their reference such as the Polish standards and publications [11, 12]. In order to illustrate the importance of the problem, completed estimation results for a meter immersed in a homogeneous field while the meter is placed parallel to the E-field force lines (right) and perpendicular to them (left) [13] are presented in Figure 8.4.

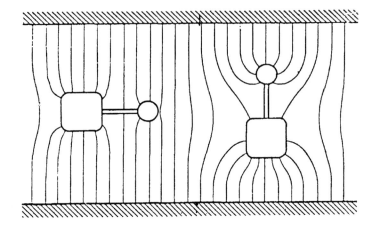

Figure 8.4. *EMF deformations caused by a measuring device [13].*

8.4 Susceptibility of the meter to external EMF

Every electric, and especially electronic device is susceptible to external EMF [14] and EMF meters are no exception. Because of their work in the near-field, in relatively high strength fields (unknown *a priori* as to value and spatial orientation), EMF hazard meters and similar devices must be immune to direct field penetration of the device and to high frequency currents induced by the field in the meter's wiring. There are multiple possibilities for field penetration: penetration through incomplete or inaccurate screens, dielectric slots on the junctions of varnished surfaces and those covered by other protective layers, penetration through metallic elements standing out of the casing (leads out of regulation potentiometers, switches, an indicator placed on the front panel) and indirect penetration by field-induced currents in the probe connection cable, other interconnection cables, power cord, etc. The first hazard meter designs worked out at the Technical University of Wroclaw almost 30 years ago (meters type MPE and MPH) were carefully screened and then their immunity to unwanted signals was experimentally proven [15]. An MPH-type meter shown in

Figure 8.5, designed for selective magnetic field measurement within the frequency range 0.1-30 MHz, illustrates the attention paid to the screening problems at that time. These efforts are now reflected in the Polish Standards PN-77/T-06581 and PN-89/T-06580/3 in the form of a requirement to test the susceptibility of EMF hazard meters.

Figure 8.5. *MPH type meter on a tripod. The meter, mounted on a standard photographic tripod for antenna rotation, controlled from the front panel, made it possible to measure three spatial components of a selected frequency fringe.*

Careful screening and filtering of any input and output of the meter may achieve the reduction of the susceptibility. These means are identical with those undertaken to limit electromagnetic interference in generally understood electromagnetic compatibility.

We should add here that the modest microelectronics, because of their relatively small sizes, may be seen as more immune to EMF. However, there may be a pitfall here. The modern meters, placed in an elegant plastic casing may be really immune for lower frequencies. But in a majority of cases, they are unprotected against microwave fields, which are becoming more and more popular. The practical results are easy to foresee and every user should be able to perform the simplest test of the meter's susceptibility (immunity). The author investigated examples of EMF meters with the electronics more sensitive to EMF than their probes, even devices manufactured by experienced firms. The investigations concluded that caution is necessary especially when using meters with plastic casings.

The problem is illustrated in Figure 8.6. The Figure shows generalized measured frequency dependence of the susceptibility to an external EMF of the V-640 type multimeter. The measurements were performed with the use of a TEM transmission line and the curve shows intensity of the applied field necessary to change the multimeter's indications by two percent. The meter investigated was self-powered with no external connection. During the measurement, the meter was almost insensitive to the position of the range selector switch, which may suggest that the most internal indicating part of the meter (amplifier, detector) was affected by the field [16].

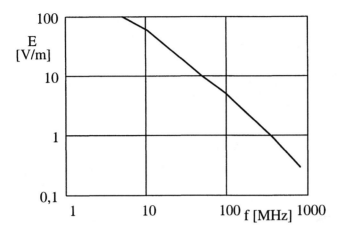

Figure 8.6. *The susceptibility of a universal multimeter type V-640 to external EMF.*

8.5. Resonant phenomena

Deformations of the measured field caused by the meter, its operator and, any conducting medium were presented above. Apart from the discussed phenomena, especially in the near-field, the resonant size objects may affect the field distribution around them more intensely than non-resonant ones. The resultant field around these objects may remarkably exceed the primary field. We will call this phenomenon field amplification by a resonant object [17]. The objects play the role of secondary passive radiators. The use of passive retranslators is well known in radio communication, a good example of their use is the Yagi-Uda antenna where only one element (radiator) is excited from a source and the other ones (reflectors and directors) are excited by the field of the radiator and because of their resonant sizes, the focusing of radiated energy is achieved. Results of the estimated magnitude of amplification in the neighborhood of a passive half-wave dipole, in the form of multipliers n(r), which should be multiplied by the intensity of the primary electric field E_0 and the magnetic field H_0 to obtain the intensity of the secondary (resultant) electric field E_s and the

176 ELECTROMAGNETIC FIELD MEASUREMENTS IN THE NEAR FIELD

magnetic one H_s in their maxima. An example of this is plotted in Figure 8.7, with the electric field near the dipole ends and the magnetic field near its center. The dipole resonant frequency was chosen as 150 MHz, however, because of the linearity of the phenomenon the presented magnitudes of amplification may be transferred to any other half-wave resonant dipole by multiplying the amplification factors read from the curves by $150/f_x$ (where: f_x = resonant frequency of the dipole in MHz).

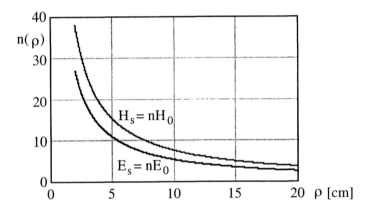

Figure 8.7. *EMF amplification of the half-wave passive dipole at 150 MHz.*

As shown in Figure 8.7, at a distance of 5 cm from the ends of the resonant passive dipole, the intensity of the electric field increases ten times whereas that of the magnetic field (at the same distance from the dipole center) increases fifteen times! Although the effect decreases with frequency, even up to frequencies of 1 GHz, it is substantial enough that it should be taken seriously.

The measured resonant frequencies of different objects applied in everyday life are shown in Table 10.

Object	f_{res} [MHz]
Table	30 – 60
Chair	70 – 100
Stool	100 – 150
Tools	300 – 500
Cutlery	300 – 600
Glasses	350 – 1000

Table 10. *Measured resonant frequencies of generally used objects.*

In order to illustrate the scale of the field deformations caused by the resonant phenomena as well as to show magnitudes of the field amplification by the resonant secondary radiators, Figure 8.8 shows the standing wave within a metal door frame at a FM station frequency around 70 MHz (the standing wave distribution was in the case used to identify the source of radiation). Figure 8.9 shows the field distribution around a metal kitchen stool at its resonant frequency near 135 MHz.

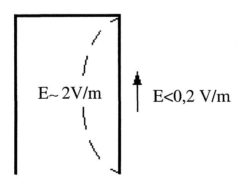

Figure 8.8. *The standing wave in a metal door frame.*

178 ELECTROMAGNETIC FIELD MEASUREMENTS IN THE NEAR FIELD

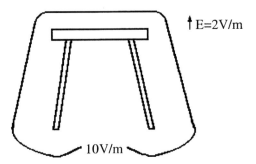

Figure 8.9. *EMF around a kitchen stool.*

In any cases illustrated in Table 10, the field amplification remarkably exceeded 20 dB at distance of 5 cm from the measured object. The effect was extremely strong in the case of eyeglasses with metal frames while portable transceivers were in use [18]. The frames not only confirm the role of the resonant phenomena and amplify the field around them, but also concentrate the field around the head of the device user. What was not taken into account in the data presented in Table 10 was the influence upon the resonant frequencies and their amplification of their proximity to any other material media. In the case of the spectacles, their resonant frequency when being worn by a person is much lower as compare to just sitting in free space; the field distribution around them, of course, is different than around a linear dipole and their ear arms form something like capacitor plates causing the above-mentioned phenomenon of field concentration inside a head.

The resonant phenomena requires not only a special care during the field measurements near them but also a modification of the protection standards in the sense of the distance at which the measurement should be performed. The example of the eyeglasses supports the point the best.

8.6. Bibliography

1. C. Dragone, "Performance and Stability of Schottky Barrier Mixers," *The Bell System Technical Journal*, Vol. 51, No. 10/1972, pp. 2169-2196.
2. N. Inoue, Y. Yasuoka, "Responsivity of Antenna-Coupled Schottky Diodes," *Infrared Physics*, Vol. 25, No. 4/1985, pp. 599-606.
3. G. Gerbi, D. Golzio, "A New EM Field Sensor for Radiation Hazard and EMI Measurements," *Proc. IEEE 1983 Intl. EMC Symp.*, Arlington, pp. 142-146.
4. K. A. Chamberlin, J. D. Morrow, R. J. Luebbers, "Frequency-Domain and Frequency-Difference, Time-Domain Solutions to a Nonlinearly-Terminated Dipole: Theory and Validation," *IEEE Trans.*, Vol. EMC-34, No. 4/1992, pp. 416-422.
5. "Geraetedokumentation Nahfeldstarkemessgeraet NFM-1" (in German), Haidenau 1986.
6. R. G. Harrison, X. Le Polozec, "Nonsquarelaw Behavior of Diode Detectors Analyzed by the Ritz-Galerkin Method," *IEEE Trans.*, Vol. MTT-42, No. 5/1994, pp. 840-846.
7. T. M. Babij, H. Trzaska, *Quarterly Reports for the NBS Grant No. NBS(G)-176* (unpublished), Wroclaw, 1975.
8. B. R. Strickland, N. F. Audeh, "Numerical Analysis Technique for Diode-Loaded Dipole Antenna," *IEEE Trans.*, Vol. EMC-35, No. 4/1993, pp. 480-484.
9. M. Kanda, "Analytical and Numerical Techniques for Analyzing an Electrically Short Dipole with a Nonlinear Load," *IEEE Trans.*, Vol. AP-28, No. 1/1980, pp. 71-78.
10. E. Grudzinski, H. Trzaska, "EMF Indicators," 1984 Annual Meeting of the BEMS, Copenhagen.
11. H. D. Bruens, H. Singer, T. Mader, "Numerical Investigations of Field Distortions Due to Sensors," *IEEE Trans.*, Vol. EMC- 35, No. 2/1993, pp. 110-115.
12. J. D. Norgard, R. M. Sega, M. Harrison, A. Pesta, M. Seifert, "Scatter-ing Effects of Electric & Magnetic Field Probes," *IEEE Trans.*, Vol. NS-36, No. 6/1989, pp. 2050-2057.

13. T. Bossert, H. Dinter, *Beurteilung der Gefaehrlichkeit starker elektromagnetischer Felder-kalibrierung und messgenauigkeit von Nahfeldsonden im Berich 30 kHz–30 MHz (in German)*, JTG Fachbericht, No.106/1988, pp. 57–62.
14. H. Cichon, H. Trzaska, "Selected Susceptibility Problems of the General Use Electronic Equipment," *Proc. 1979 Intl. EMC Symp.*, Rotterdam, pp. 123–126.
15. T. M. Babij, H. Trzaska, "Selective EMF Meters'" (in Polish) *IME Publications of the Technical University of Wroclaw*, No. 1/1970, pp. 55–66.
16. H. Cichon, H. Trzaska, "Susceptibility Problems of Home Entertain-ment Electronic Devices," *Proc. 1981 Intl. EMC Symp.*, Zurich, pp. 295–299.
17. H. Trzaska, "Resonant Phenomena and Their Role in Dosimetry and Protection Standards," Seventeenth Annual Meeting of the BEMS, Boston 1995, p. 188.
18. H. Trzaska, "EMF Near Passive Secondary Radiators," USNC/URSI Radio Science Meeting, Newport, CA, p. 59.

9

Photonic EMF Measurements

Until now, the dominant technique for EMF measurement was to use an antenna (mainly a dipole or a loop) loaded with a detector (diode or, more rarely, thermocouple) and transfer the DC voltage from the probe to an indicator (in the case of the most popular designs of two-piece meters) through a high resistance (transparent) transmission line (Figure 9.1a). The technique's most important inconvenience is the vanishing of phase information that is (sometimes) indispensable; similarly, the spectral information is lost as well. Although the latter is usually unnecessary, especially where wideband measurements are of concern, we will see in further considerations that it is possible and advantageous to perform wideband measurements using spectral information as well.

182 ELECTROMAGNETIC FIELD MEASUREMENTS IN THE NEAR FIELD

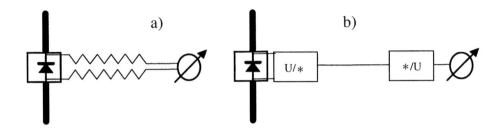

Figure 9.1. *EMF meters: a) traditional design with the resistive transmission line, b) design using a photonic modulator.*

The use of photonic media amongst all considered media of information transmission seems to be, in our field of involvement, the most promising one. Only recently have proposals been worked out for photonic field meters and the first designs are being investigated and even applied under laboratory conditions. This technique is often applied for data transmission and for remote reading of measured values. An example of this meter design is shown in Figure 9.1b.

The advantages of this concept are illustrated by the EMF meter designed at the Technical University of Wroclaw and used mainly for EMF measurements in proximity to transmitting antennas. The measurements were and are performed for experimental verification of theoretical hypotheses in order to evaluate exposure of the nearest inhabited areas on the spatial EMF distribution around transmitting centers when modifications or reconstruction were planned. Similar measures are done when new buildings are planned near transmitting centers. According to the Polish environment protection standards, in these situations the theoretical estimation of the exposure, in relation to the limits provided by the standards, is an initial requirement for investment approval. Then, after the investment is completed, the estimations should be confirmed by measurements.

For this purpose, a suitably modified EMF meter, equipped with a probe adequate for the selected frequency range and to the measured component of the field, was hung under a weather balloon. The output voltage from the meter was converted into an optical signal and transmitted to a demodulating device, on the ground through an optical fiber and then to a recorder (Figure 9.2) [1, 2].

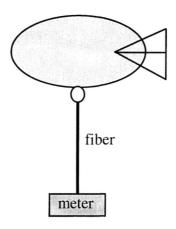

Figure 9.2. *Spatial EMF distribution measurement using a captive balloon.*

Contrary to other wire or wireless transmitting media, the presented solution is absolutely unsusceptible to external interference and, especially important, is insulated against unwanted influences of the measured field. Unwanted influences on the measured field are presently a source of many problems, which are the subject of involvement of electromagnetic compatibility, and in particular the remarkable sensitivity of any type of electric or electronic device to an external EMF or to voltages induced by the field in the device's wiring and interconnection cables, as mentioned in Chapter 8.4. The optical fiber used for the data transmission (used simultaneously as a captive line of the balloon) does not cause any disturbances of the measured field. The flying part of the measuring system is insulated galvanically against the meter on the ground; which ensures that measuring team is absolutely safe from the possibility of the shock by high frequency currents, power line frequency currents, or static charges. As a result, measurements can be performed in close proximity to big power antennas, their guy wires, towers, overhead power lines, and other charged devices.

Although this construction has many advantages (some of which are listed above) it is relatively primitive since the optical fiber is used only to make it possible to get remote readings of the measured data. As already mentioned, similar designs (with considerably reduced transmission distances and other requirements) are presently offered by many manufacturers in the form of so-called repeaters that could be connected to EMF meters and indicators. Later in this chapter we will discuss the possibility of using of photonic elements as sensors (transducers) of the measured field. The advantages of using the photonic technique for data transmission as well are, in some sense, by-products of the method.

Introductory comments:

A. There is abounding literature devoted to the field of opto-electronic elements' use as sensors for a variety of physical quantity measurements. A number of publications are concerned with the opto-electronic transducers use for EMF measurement [3, 4]. Because of this, we won't present particular solutions in detail, but we will concern ourselves with a few of them that are especially useful for our applications in order to illustrate the tools and the methods that in all likelihood will dominate in the near future.

B. The principle of a photonic transducer may be conveyed to the modulation of an optical beam and the subject of the modulation may be the signal's phase (which is a function of the light's velocity of propagation in an electro-optic medium), its frequency, amplitude or polarization. The type of the modulation as well as that of the electro-optic crystal are selected in such a way as to obtain the device's maximal sensitivity. Lasers are usually used as coherent light sources, with monomode semiconducting lasers being the most common.

C. While current literature discusses the sensitivities of photonic meters exceeding that obtainable using traditional techniques, sensitivity is still the Achilles heel of the photonic technique and its increase is the subject of current technological efforts (the development of electro-optical materials with sensitivity exceeding materials that were used until now) and the technical ones (the development of electro-optic systems and electronic circuits that will allow an increase in sensitivity while maintaining the necessary stability and reasonable prices).

D. In general, in the discussed application, two approaches are used:

1. A direct interaction of the measured field onto the electro-optic crystal.
2. A voltage (or rather an emf) induced by the field in an auxiliary antenna (playing the role of the measured field concentrator) impressed to the modulator.

The latter is usually used in order to increase the sensitivity of the meter by the application of relatively large-size antennas. Although in the former approach, large size sensors (for the sensitivity increase) may be applied as well, it results in a reduction of the system's stability. In the latter, analyses of the measuring band (including the necessity of the RC frequency response shaping filters use) are identical to that presented in Chapter 4.2 [5]. In the former, there is no lower corner frequency (it also allows measurement of the static fields) while the upper band pass limit depends only upon the properties of the electro-optic material applied in the modulator and its size.

E. Apart from the technological differences, the factors limiting measurement accuracy are, in the case of the photonic probes, similar or identical to those presented in previous chapters. They may be related to the frequency response of the probe, which was mentioned above, its sizes (not taking into considerations interaction of a source upon the properties of the probe in the version with no antenna), deformation of the spherical directional pattern, susceptibility to the temperature alternations and factors specific to the photonic technique. For instance, instability of the power generated by a laser, optical power losses, modulator heating, and others.

9.1. The photonic EMF probe

Contrary to the design shown in Figure 1b, in a photonic sensor (without detection) the measured EMF with no changes or modifications to its envelope is applied for modulation of a coherent light beam (Figure 9.3). As mentioned, the modulator may be affected directly by the measured field or (for the purpose of sensitivity increase) an antenna (of electric or magnetic type) may be additionally applied.

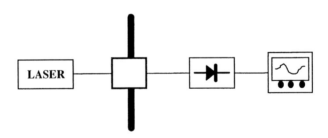

Figure 9.3. *Block diagram of a photonic EMF meter.*

In photonic EMF sensors, it is possible to modulate the beam in amplitude, phase or polarization. Although the final result of the procedure is always amplitude modulation, direct amplitude modulation is rarely applied because its relatively small sensitivity and, as a result, its unacceptable signal to noise ratio. Block diagrams of a photonic EMF meter using polarization modulation and phase modulation are shown in Figure 9.4.

Both solutions shown in Figure 9.4 have similar parameters. Thus, for illustration we will discuss in more detail one of them — the probe with the Mach-Zehnder interferometer. The laser beam is split in the divider into two parts (approximately equal to one another). One of them makes the reference beam, whose phase (propagation velocity) should be independent of the measured field, while the other, due to alternations of the modulator's permittivity as a function of its application to a field and as a result of the phase velocity alternations (modulation) of the optic ray, is phase modulated. Then the beams are interfered in the adder, which results in the amplitude modulation of the primary light beam at the output of the interferometer.

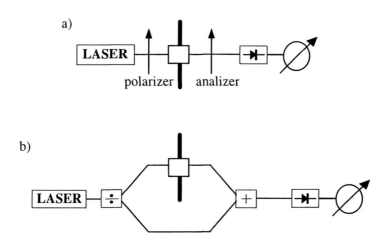

Figure 9.4. *Two types of photonic EMF meters: a) polarimetric, and b) interferometric.*

The output beam is directed to a detector and then a measuring device analyzes the signal. The signal's modulation depth, in a certain range, is linearly dependent on the amplitude of the measured field while the frequency of the signal (including its envelope) is exactly equal to the modulating frequency. The statement concerns every frequency fringe of the signal that is lead to the modulator from the antenna (measured field). Because of the modulator linearity, the unwanted products of intermodulation, a distinctive feature of active antennas, should not appear at the detector's output. The latter well illustrates the role of the presented device not only for measurement purposes but as an active receiving antenna as well. In the device, the intensity of the optical signal at the detector (I) is given by:

$$I = \frac{I_0}{2}\left[1 + \sin\left(\pi \frac{V}{V_{\lambda/2}} + \varphi_0\right)\right] \quad (9.1)$$

where I_0 = light beam intensity at the output of the modulator
V = modulating voltage
$V_{1/2}$ = the voltage causing phase difference in the modulator in p
φ_0 = initial phase

The introductory phase results from an inaccuracy of the optical tracts manufacturing and alignment. Because of its importance for the adjustment of the modulator's working point, on one hand the modulator set-up requires precise work and on the other hand requires the possibility of controlling the working point without the necessity mechanical interference into the device. Figure 9.5 illustrates the role of the introductory phase calculated using formula (9.1), and the output signal of the interferometer for $\varphi_0 = 0$ and for $\varphi_0 = p/2$ [6].

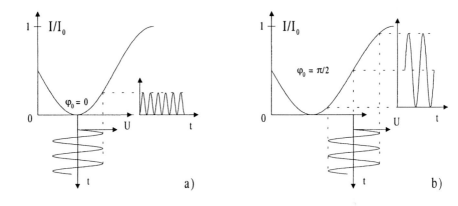

Figure 9.5. *Dynamic characteristics of an interferometer a)* $\varphi_0 = 0$ *and b)* $\varphi_0 = p/2$.

Notice here that the solution shown in Figure 9.5b is more profitable both because of its relatively large range of the linear part of the dynamic characteristic and its higher sensitivity. In both cases the characteristics are periodic and as a result, if the modulator input signal exceeds a certain value, the output voltage of the detector not only does not increase but it decreases.

9.2. Frequency response of the probe

Although we have ascertained that there are similarities between the analyses of the frequency response of a traditional probe and for that of a photonic one, some features of the response, especially those distinctive to the photonic probe, which may be useful for emphasizing its advantages, require a more accurate approach and focus of our attention on differences rather than on similarities.

190 ELECTROMAGNETIC FIELD MEASUREMENTS IN THE NEAR FIELD

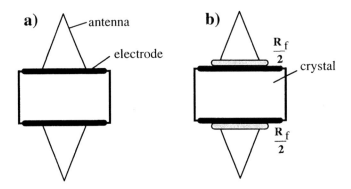

Figure 9.6. *Structure of a photonic EMF probe: a) without RC filter, b) with the filter.*

In order to shorten our discussion, we will refer to Chapter 4 and we will assume, without additional explanation that, as in the case of the traditional probes, for photonic probes in the range of the highest frequencies the measuring band limitation is indispensable here as well. Figure 9.6 shows the structure of a probe with an optical modulator without an RC low-pass filter and with such filter.

Figure 9.7 illustrates an equivalent network of the probe with the low-pass RC filter.

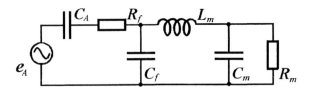

Figure 9.7. *Equivalent network of the probe with the RC filter.*

Using certain modifications of the equivalent network shown in Figure 9.7, in relation to those applied in Figure 4.2, the transmittance of the probe shown in the former is given by a formula identical to (4.12) while we assumed $L_f = 0$. Let's direct our attention to the specific differences in the transmittance's frequency run for both the traditional probe and the photonic one.

9.2.1. The lower corner frequency

The lower corner frequency f_l of the photonic EMF probe, at which its transmittance diminishes by 3 dB in relation to the value within the measuring band, using the indications shown in Figure 9.7, may be defined in the form:

$$f_l = \frac{1}{2\pi R_m (C_A + C_f + C_m)} \quad (9.2)$$

Formula (9.2) is identical to formula (4.15), however, it requires some comments.

a) If the measuring antenna is loaded with a detection diode, as a result of the diode's non-linear properties, both its internal capacitance and the detector equivalent resistance are functions of current's intensity flowing through the diode. Thus, both the lower corner frequency and to a lesser degree, the transmittance within the measuring band are dependent upon the measured field strength, which has already been signalized in Chapter 8. In the case of the photonic probe with the phase modulation, the modulator equivalent capacitance C_m also varies as a function of the applied voltage (EMF), because the modulation is realized through ε_r variations in the voltage function, but the dependence is much less than that of the diode detectors. On the other hand, the modulator equivalent resistance R_m is voltage independent. Thus, the stability of the lower corner frequency and the absolute magnitude of the transmittance within the measuring band are, in the photonic sensors, above those of traditional ones. The comparison is more advantageous for the photonic probe in the aspect of its thermal stability; although it is sensitive, to a certain degree, to the temperature alternations as well, but in its optical part.

b) The acquisition of a low f_l in the traditional probe requires careful selection of the type and, even, specimen of the detection diode to obtain a high value of R_d. In the photonic probes, the high values of R_m are as the rule, which makes it possible to achieve a lower corner frequency well below that of traditional probes.

c) We have already mentioned that there exists the possibility of directly using an optical modulator in the role of the field probe. In this case, the antenna capacitance C_A does not exist. If we assume in formula (9.2) $C_A \to \infty$, we will have $f_l \to 0$! The sensitivity of such a probe won't be very high, however, the field measurement starting from the static fields is so attractive that this possibility is the subject of numerous investigations. We should notice here that the statement may refer to both the electric and magnetic field. Although in the magnetic field sensor, the possibility is doable only in the version using a modulator with a magneto-optic crystal. In the version with a loop antenna and an electro-optic modulator, the lower corner frequency is similar to that of probes with loop antennas, which were discussed in Chapter 5. A similar result, with a much narrower frequency range may be reached using a hall-cell. But in the case of an electric field probe, the possibility is not assured by any other measuring method currently known.

9.2.2. The upper corner frequency

The upper corner frequency f_u of the probe shown in Figure 9.7 is given by:

$$f_u = \frac{C_A + C_f + C_m}{2\pi R_f C_A (C_f + C_m)} \tag{9.3}$$

The structure of the photonic probe shown in Figure 9.6 allows the supposition that the modulator inductance L_m is much lower in comparison to the parasitic inductances of traditional probes, even if they use thin-film hybrid technology [7]. As a result of this, the filter capacitance C_f and the modulator capacitance C_m should be understood as a single element, which we will call the resultant capacitance of the modulator C'_m. Consequently, the self resonance of the modulator will appear at a frequency much higher than similar resonances in traditional probes and an adequate time constant of the $R_f C_f$ filter can be smaller in relation to the considerations of Chapter 4. In the no-antenna probes, the filter limiting the upper corner frequency is not necessary (if, at least, its realization is possible at all). The measuring band of the probe is limited in the highest frequency range by the maximal sizes of the probe and defined, for example, by the permissible value of the errors δ_{1E} and δ_{2E}, which were estimated in Chapter 4, or by analogous limitations of the electrical and the geometrical sizes of the magnetic field probe, which were considered in Chapter 5. In the latter case, the antenna effect does not exist in relation to the no-antenna probe with no regard for the sensitivity of the photonic electric field probe to the magnetic field and *vice versa*. It is worth remembering that in the case where the upper frequencies are artificially unlimited, the run of the transmittance may be, in this range, arbitrary and, as a result, measuring errors (in case of signal presence within the range during measurements) may be arbitrary as well.

Apart from factors discussed limiting widebandness of the photonic probes, attention must be focused on a very important factor — the simultaneous appearance of piezoelectric phenomena in electro- and magneto-optic crystals. Then when the modulating frequencies are near those of the piezoelectric resonances, a remarkable increase of the modulation efficiency resulting from coincidence of the two phenomena will appear. These resonances do not exclude the possibility using such a crystal in an optical modulator at frequencies above the first piezoelectric resonance, however, the frequency response of the modulator above the frequency will have a comblike character. Meaning that at the following harmonics of the basic resonant frequency and at harmonic frequencies of any other crystal oscillation modes there will appear strong and narrow maxima. Sometimes in metrological applications it is enough to precisely know all these frequencies and, as a result, the shape of the frequency response of the modulator (probe), then they may be taken into account during computer analysis of the measurement results. However, the resonances strongly depend upon the temperature, which makes the analysis inaccurate. It results in the trend to use EMF probes with a flat frequency response within the measuring band, then the upper part of the response should be eliminated artificially (for instance by the use of the presented RC low-pass filters). In a designed and investigated model of the modulator using an $LiNbO_3$ crystal, the resonant effects are very strong; its sensitivity at the resonant frequencies is several times that of the non-resonant frequencies. The phenomenon is well illustrated by the measured frequency response of the hybrid modulator (Figure 9.8) [7].

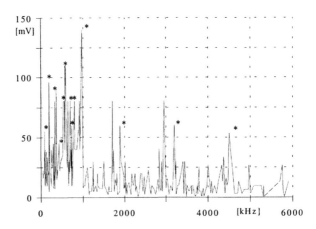

Figure 9.8. *Measured frequency response of $LiNbO_3$ modulator.*

For the frequency band where the resonances appear, the first resonant frequency should be artificially eliminated and, as a result, the upper corner frequency of the probe may be estimated as:

$$f_u \approx \frac{v}{2d} \qquad (9.4)$$

where v = velocity of the acoustic wave propagation in the crystal (usually 4–5 km/s)
    d = transversal size of the modulator

9.3. Sensitivity of the photonic probe

The detector output voltage V_d in the Mach-Zehnder interferometer with an optimally selected working point, i.e., $\varphi_0 = \pi/2$, for the medium frequency range, is given by:

$$V_d = \frac{CI}{2}\left(1 + \cos\pi\frac{V}{V_{1/2}}\right) \qquad (9.5)$$

where I = the light beam intensity at the detector
    C = constant, it reflects the efficiency of the detection including attenuation losses in the optical track
    V = modulating voltage:

$$V = Eh_{eff}T = Eh_{eff}\frac{C_A}{C_A + C_m} \qquad (9.6)$$

If we assume the typical magnitudes of applied power, reasonably measurable values of V, and the parameters of the modulator, then we will have the sensitivity of the device, i.e., the minimal EMF intensity that could be measured with its use. The estimation, completed for the most often applied modulator with $LiNbO_3$ crystal, gives a result somewhat below 100 V/m.

The estimation results are approximately two orders of magnitude below those experimentally confirmed for similarly sized traditional probes. Thus, taking into account other advantages of the photonic probes it is necessary to initiate further work to increase its sensitivity. The most evident approach here, as mentioned above, is the development and application of new crystals having lower values of the $V_{\lambda/2}$ voltage. These approaches have been undertaken and there are already known crystals that allow the better sensitivities as compared to the lithium niobiate, although their availability is currently very limited. Below we will present two solutions that allow an increase in the sensitivity by using specific circuitry, which may be applied to an arbitrary crystal type.

9.3.1. The modulator with multiple access

The phase change δ observed at the interferometer's output is the result of the difference between the refractive indexes in the reference track of the interferometer and that in the modulated arm and is caused by the applied voltage that we will write in the form:

$$\delta = \frac{2\pi}{\lambda_o} (n_m - n_0) L \qquad (9.7)$$

where L = the modulators length
 n_0 = refractive index without modulation
 n_m = refractive index while a modulating voltage is applied

As far as the influence of the refractive index alternations on the phase difference reflects the role of the crystal (its sensitivity) applied in the modulator; the length of the modulator that appears in formula (9.7), suggests the use of large sized modulators in order to obtain the probe's required sensitivity. As already shown in Chapters 4 and 5, the sizes are rigorously limited, primarily as a result of the measured field averaging, particularly while the measurements are performed at distances to a source comparable to or less than the sizes of the probe. The perfect probe for our purposes would be a zero-dimensional probe. In the case of the magnetic field a good approximation of the ideal is a probe with a hall-cell or a magneto-diode, but their weaknesses radically limit the use of these transducers as compared to loop antennas. Thus, the increase of the modulator's size in the photonic probe, although it allows a certain increase of sensitivity, leads to the limitation of its use especially in close proximity to the radiation sources and requires some limitation of the measuring frequency band since, apart from the resonant phenomena, the modulator works most effectively when changes of the measured field phase along the modulator are small.

An artificial increase of the modulator's length, without necessary changes of its geometrical dimensions, makes it possible for multiple passages of the modulated light beam through the modulator. Figure 9.9 illustrates three variants of this solution.

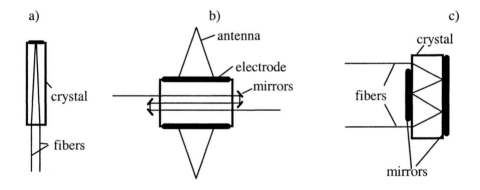

Figure 9.9. *The multiple access modulators.*

Figure 9.9a shows the simplest solution of the double passage of the light beam with its reflection at one of the ends of the modulator and its withdrawal from the modulator through a separate (if necessary) fiber optic from the entry side. Because of remarkable attenuation of the light beam in electro-optical crystals, the interfering influence of the multiply reflected rays upon the resultant phase may be neglected. Figure 9.9b shows a modulator with multiple passages in which (contrary to the other two solutions) the light beam is parallel to the optical axis of the crystal. This method obtains relatively good independence from the multipath propagation within the crystal as well as the most advantageous conditions of beam modulation. A 'multiplied' version of the solution from Figure 9.9a is shown in Figure 9.9c. Two opposite sides of the modulator were equipped with reflecting surfaces (mirrors). The input and the output optical fibers are placed in the corners, free of the reflecting metalization, and aligned in such a way as to obtain the required number of beam passages through the modulator [8].

9.3.2. The heterodyne sensor

Two lasers, optically excited by a third one (in order to assure the galvanic insulation of the probe against other parts of the measuring device and to feed the probe throughout the 'transparent' line) create the heterodyne EMF probe (Figure 9.10) [9].

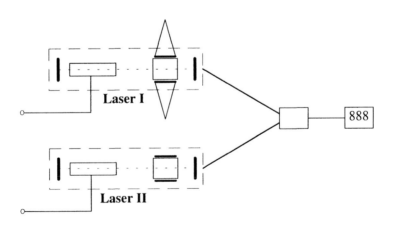

Figure 9.10. *Block diagram of the photonic heterodyne probe.*

One of the lasers is modulated in frequency by a voltage fed from the measuring antenna to an electro-optic crystal immersed inside the laser's resonator. The other laser plays the role of the heterodyne and its modulator is used for the possible mean value of the differential frequency stabilization that allows for compensation of its instabilities. Frequencies of both lasers are mixed and the differential signal, after detection, is lead to an indicating device.

This concept seems to be extremely attractive because of the possibility of achieving sensitivities exceeding those of the traditional probes. The main problem here is the stabilization of the devices work. To improve thermal stability, both lasers are identical is possible and placed close to one another in the same external conditions (temperature, pressure, humidity, vibration, acoustic noise, and light). The results of thermal creep, the laser-exciting power variations, and other factors affecting stability are compensated using a heterodyne modulator. It is also applied for setting the working point (differential frequency) of the system and to introduce a feedback. The modulator is fed by a separate optical fiber. We have not mentioned that the measured frequency range in the solution may be more rigorously limited by difficulties with wideband FM detection than by the above-discussed phenomena

9.4. Linearity of the detector

As can be seen from the shape of the dynamic characteristics of the interferometer's detector (and similar to the polarimetric system) shown in Figure 9.5b, the linear part of the characteristics is limited only to the range in which the function sine is represented with required accuracy by the first term of its expansion in the power series. Moreover, the output signal of the interferometer is a periodical function and when the modulating voltage exceeds $\pm V_{\lambda/2}/2$, the output voltage decreases. Thus, the linear measuring range of the photonic probe (modulator) is relatively narrow, which causes problems with required sensitivity achievement on one hand and problems with linearity of the detector and the limitation of the maximal values of the modulating voltage (and the measured field) on the other. The sensitivity improvement methods were proposed above. In order to linearize the device and to exclude unwanted consequences of the detector's periodical properties (i.e., to extend the range of measured values) a compensational concept of the meter has been devised (Figure 9.11).

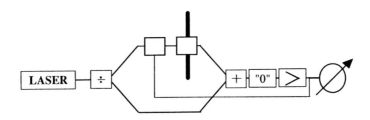

Figure 9.11. *Compensational EMF measurement.*

The modulator presented in Figure 9.9.b was designed and completed in a manner that makes it possible to modulate all the transitions with one common voltage or separate excitation of all of them. One of the pathways may be used for the negative feedback. Thus, an output voltage of an amplifier excited by a detector feeds it. The voltage that is simultaneously fed to an indicator as a measure of the investigated field strength, is equal, in its magnitude, to the modulating one while its phase is opposite. This way the compensation of the modulating signal is achieved. The compensational approach makes it possible to suppress the problems with the modulator linearity and makes it possible to measure much larger field strength [10]. A disadvantage of the device is a certain limitation of its measuring frequency band resulting from the necessity of using in the feedback, fast amplifiers with high amplification and large dynamic range (their dynamics limit those of the measured field). With no regard to this, the presented concept may be helpful even in a limited frequency range when wide dynamic range of measured field is required.

An extension of the concept, based on the use of additional modulating electrodes in the reference track of the interferometer, permits use of the interferometer electric tuning and feedback for stabilizing its working point [7].

9.5. Synthesis of the spherical directional pattern

Because of the presented considerations, the output voltage of an optical detector precisely reflects (as to its amplitude and the phase) the spectrum of the measured field (signal). If frequency fringes of different polarizations appear in the spectrum and if we want to take them all into account, while interpretation of the measured field is prepared, the use of an omnidirectional probe is indispensable. Let's consider several problems specific to the spherical pattern synthesis of the photonic probes.

Using the methods presented in Chapter 7 and taking into account linearity of the photonic probe it is possible to design an omnidirectional probe composed of three identical and linearly independent probes of sinusoidal pattern and then summing squares of their output voltages. This method is evident and efficient, however, it requires use of a relatively complex measuring system. An essential question arises here: the result of summation of the squared spatial components of spherically polarized signals is a constant value. In Chapter 7 it was, with no regard as to how the measured signal was interpreted, a DC component. The matter was in the DC component as far as it was the subject of the measurement, i.e., the component was proportional to the intensity of the measured E or H field component. Meanwhile, in the considered case the (optical) detector is loaded with a spectrum analyzer, a selective high frequency millivoltmeter, an oscilloscope, or another device for alternating voltages' measurement.

The constant value of the output voltage, provided in the spherical polarized fields measurement, is the result of filtration of the output signal as well as large time constants of the detectors. In the measured field, in an arbitrary moment of time, there exists one and only one linearly polarized vector of the E-field and one of H-field. We should notice that in the optical detector there are no filtering (averaging) elements and as a result, it is 'fast' and the output voltage of the detector is an AC voltage, proportional to the instantaneous field strength that represents the amplitude, phase and frequency of the field.

The construction of an omnidirectional probe using traditional technology (an antenna loaded with a diode) was relatively simple as it was assumed (with an accepted accuracy and within limited dynamic range) that the characteristic of the diode is square-law and the summation of the output voltages of three mutually perpendicular probes was enough to obtain the required omnidirectional properties. In the case of photonic sensors, a similar approach is possible. It is necessary to use probes (modulators) whose working point was selected for $\varphi_0 = 0$ (Figure 9.5a) to achieve this, however. Their output voltage is proportional to the square of the modulating voltage. Thus, by way of parallel connection of three identical probes (Figure 9.12), positioned in three mutually perpendicular directions and sensitive to three different spatial components of the measured field, it is possible to synthesize, in a relatively simple manner, the omnidirectional pattern. Similar results may be obtained when the probes are in serial connection, however, in the latter case their working point must be shifted by π in relation to the previous one. Another version of the solution shown in Figure 9.12 may be realized by replacing the order of the summation and the detection. Assuming the detector is a linear device, we will have exactly the same result, its additional advantage will be the possibility of independent measurement of the spatial components of the investigated field. Unfortunately, as may be seen from the curves shown in Figures 9.5a and b, the simplification of the omnidirectional probe construction was obtained at the expense of remarkable reduction in its sensitivity. But, even here, the attractiveness of the simpler construction may be sometime crucial [11].

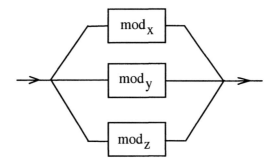

Figure 9.12. *The omnidirectional photonic probe.*

9.6. The future meter

Current-day meters used for labor safety and protection of the general public, fulfill substantially formulated demands, especially in specific applications for surveying and monitoring services. Apart from the disadvantages of the meters, presented in previous chapters, their most important imperfection is the misleading possibility of identical indications of the meter when fields of different parameters (frequency spectrum, modulation type, polarization) are measured and not precisely defined magnitudes characterizing the field, which should be measured. These unclear measurement conditions may lead to non-equivalent investigation results not only under laboratory conditions (where, at least, parameters of the applied field are, in the majority of cases, results of planned selection and the field parameters are known beforehand with some degree of accuracy) but also those in epidemiological studies, for instance. This leads to a total impossibility of comparing the exposure data of selected populations when different measuring teams using a variety of measurement equipment perform the measurements. The problem becomes particularly critical when steps are taken toward a complex evaluation and comparison of the professional or/and nonprofessional exposure of selected groups of people and those of everybody involved.

It seems that the most sophisticated measuring demands are fulfilled by the meter equipped with wideband active antennas connected to a spectrum analyzer and then to a computer. Currently, applied active antennas require their connection with a measuring system via a coaxial cable, which is non-permissible when measurements are performed in the near-field. Using techniques discussed under the photonic approach, it is possible to construct the set whose block diagram is shown in Figure 9.13.

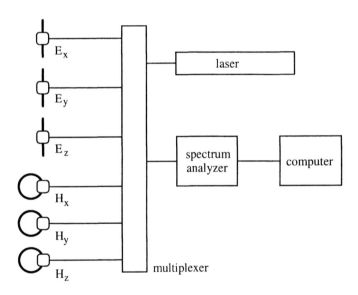

Figure 9.13. *Block diagram of the 'future meter.'*

Three mutually perpendicular magnetic field probes and a similar set of the electric field probes (the solution seems to be more profitable as compared to that using a set of multiply loaded loop antennas, because of full independence of the measurement from the E- and H-field spatial components) are through a sampling and multiplexing system connected to a spectrum analyzer controlled by a computer. Every frequency fringe spatial component is being measured by the automatically controlled spectrum analyzer, which allows complex analysis of the field. Contrary to the meters available on the market, whose frequency response is matched to specific national or other standards, these meters are designed to use entirely within the area where the standard is valid and the use of the data processing allows an arbitrary interpretation of the measurement results. A flexible selection of the software option becomes decisive instead of a meter frequency response set by a manufacturer with no possibility of changing it during meter use. The software flexibility leads to the possibility of using antennas (probes) of an arbitrary, but known frequency response. As a result, management of the measured results is simple, analysis in an arbitrarily selected aspect is possible, and standardization of the data interpretation methodology does not create problems, nor does archiving of the selected measurement results. The presented option, apart from the possibilities described, allows observations of the temporal variations of the investigated field and may play the role of a dosimeter, i.e., a device counting the product of time and the measured value (dose) [12].

The construction of such device is already possible and the first steps toward its completion have been done. Although its price is comparatively high thus far and, as a result, its applicability limited to very specified measurements, such as those in a laboratory, where the maximal quantity of information about investigated field is desired, in the future, due to an expected increase of development and accessibility of the photonic technologies followed by the decrease of the prices, the presented device may become a fundamental tool of monitoring services as well.

It may be supposed that these meters will be available in different versions, some even more advanced. For instance they might be equipped with more probes connected to one measuring device for spatial field distribution investigations or for gradient studies. Of course, versions destined for general use should be simplified in their construction, measuring possibilities and maintenance and, as a result, less expensive.

As it has already been mentioned, measuring devices used today, although loaded with many disadvantages and inconveniences, may satisfy almost all current measurement requirements. A significant part of these negative features won't be removed from the most intelligent solutions, as well. If presently, the practical ability to use relatively simple meters creates so many problems, especially for non-technicians, it should be expected and accepted that the use of devices offering much wider measuring possibilities, and as a result much more sophisticated ones, will force their users to become more educated.

9.7. Bibliography

1. E. Grudzinski, R. Kinda, J. Poreba, Z. Siwek, "Remote EMF Strength Measurements Above Earth's Surface," (in Polish), *Proc. Natl. Telecomm. Symp.*, KST-89, pp. 330–335.
2. E. Grudzinski, H. Trzaska, "General Public Protection Against Electromagnetic Radiation," *Proc. Intl. EMC Symp.*, Nagoya 1989, pp. 742–746.
3. M. L. Van Blaricum, "Photonic Systems for Antenna Applications," *IEEE AP Magazine*, Vol. 36, No. 5, Oct. 1994, pp. 30–38.
4. M. Kanda, "Optically Sensed EM-Field Probes for Pulsed Fields," *Proc. of the IEEE*, Vol. 80, No. 1, Jan. 1992, pp. 209–214.
5. T. Babij, H. Trzaska, "Properties of Wideband Magnetic Field Probes," *Proc. 1976 IEEE Intl. EMC Symp.*, Washington, D.C., pp. 375–380.
6. E. R. Mustiel, W. N. Parygin, *The Light Modulation Methods*, (in Russian), Svyazizdat, Moscow, 1973.
7. P. Bienkowski, H. Trzaska, "Frequency Limitations in Photonic EMF Probes," *Proc. 1997 Intl. EMC Symp.*, Zurich, pp. 603-606.
8. H. Trzaska, "Photonic Electromagnetic Field Probes," *Proc. EMC 1996 Roma Intl. Symp.*, pp. 221-226.

9. P. Bienkowski, H. Trzaska, "The New Approach to the Photonic EMF Measurements," *Proc. Intl. EMC Symp.*, Wroclaw, 1996, pp. 347-350.
10. P. Bienkowski, H. Trzaska, "New EMF Photonic Sensors," USNC/URSI Meeting, Boulder, CO, 1996.
11. S. Diba, H. Trzaska, "Isotropic Receive Pattern of an Optical EMF Probe Based Upon a Mach-Zehnder Interferometer," *IEEE Trans.*, Vol. EMC-39, No.1/1997, pp. 61-63.
12. P. Bienkowski, H. Trzaska, "Photonic EMF Measurements," *25th General Assembly of the URSI*, Lille, France 1996, Abstract, p. 581.

10

Final comments

This book discussed the formal and technical necessities and EMF measurement possibilities in the near-field, especially for labor safety and environment protection purposes. Field measurement technical problems in the near-field are similar to or identical to those in the more widely understood area of electromagnetic compatibility. Some differences between the far and near-field were emphasized, especially those essential to measurements performed in the near-field. We presented EMF measurement methods in the near-field and as were factors limiting accuracy of the electric and magnetic field and the power density measurement in the near-field using small electric and geometric size antennas. The influence of external factors, which are important for the accuracy estimations, was outlined and current measurement techniques, based mainly upon the development of the photonic techniques, were briefly presented.

The considerations presented here allow us to formulate three summarizing questions:

1. The presented estimations are analytical in nature. Is it possible to perform a synthesis that would allow the final, synthetic conclusion regarding measurement accuracy?

2. What is the relation between the presented material and present legal regulations and how should future regulations be directed?

3. What are the perspectives of the field development?

We will try to briefly present the author's views relating to the questions above and as well as doubts surrounding these questions. These opinions have resulted from both the character of the presented work and from the author's knowledge and experience. This book was prepared as an introductory selection, summarizing and assembling the present knowledge in the field and was provided as a background for broader discussion that may take place within the context of one or more competent bodies. The author welcomes elaboration on the present status of the near-field EMF metrology based on this book.

Estimation of the entire measurement accuracy, taking into consideration the specificity of the measurements for labor safety and environmental protection purposes, seems to be both impossible and unnecessary. The initially undefined object of the measurements may substantiate the impossibility. Remember that in many cases, the primary aim is to reveal and to identify the source of radiation and then to measure the generated field. It is only when the location of the primary source and other objects having an effect on the source-generated field have been identified, that precise considerations of the measurement accuracy can be completed. This can only be done if the propagation's geometry is well known and can only be done for the defined point in space where the probe is placed. The purpose of our measurements is to accurately determine the measured quantity results. In our case we are particularly interested in finding maximal values of time and space. However, these values may be characterized by rapid alternations, which in many cases may well exceed the applied measuring device's inaccuracy. For example, in the process of dielectric welding, the field intensity near the device changes two to five times during a single welding cycle whose duration does not exceed several seconds. When during the cycle a person will try to take a measurement is impossible to predict, although probably the measured magnitude will be read near the end of the cycle when the field becomes almost constant. However it is unknown if the value was maximal, minimal or something in between. All of it is done using a meter with an estimated measuring error of say, 10 percent. The procedure may be described with the use of a proverb that translates into English as: 'breaking a butterfly on the wheel.' It does not mean that the measurement inaccuracy is arbitrary.

Our discussion of the factors limiting the accuracy shows that the most essential factors are the distance between a source and a measuring probe, the size of the measuring probe and the thermal effects. The presence of any objects that cause deformation of the measured field within the measurement area must be included in our accuracy considerations. We should understand the effects of their possible spatial reconfiguration due to technological, operational or other reasons. A set of separate problems is created by the pulsed fields measurement. In order to avoid the problems mentioned, as well as other ones, accurate study of the applied equipment's parameters, its advantages and disadvantages and its correct use is required. It may be said, with some approximation, that quite good measurements may be performed using a relatively poor device if its parameters as well as the measurement conditions are known and understood. A separate and previously unmentioned factor of measuring device accuracy is its calibration accuracy. Accuracy usually does not exceed 0.5 to 1 dB.

The legal acts in this field play a very important role because they are valid. However, the majority of national standards, and in particular, international regulations and recommendations, are rigid. Their amendment, revision or transition to the newest trends and knowledge can be difficult or sometimes, almost impossible (as in the case of Polish standards). An outstanding example is here the ANSI standard, which is periodically reviewed and revised. This standard was born from international regulations (WHO, ITU, IEC,

URSI, EC) and apart from the many different organizations involved or, maybe because of this, it is difficult to characterize as noted by Dr. Kunsch [1.18]. It is the author's hope that it should be possible to establish one competent international body to prepare a reasonable proposal. The community in the field is small (as compared to many other fields) and it is grouped more or less officially around the Bioelectromagnetics Society (BEMS) and the European Bioelectromagnetics Association (EBEA). The members of the organizations dominate, in different combinations, in the regulatory bodies mentioned.

The author, as a technician, does not reserve any right to decide what threshold limits should be accepted for future standards. His opinion should be auxiliary to the decisions of medical personnel, physicists, and engineers. Of course, participation of a technician in any working group or researchers' team is indispensable just as in biology; the biologists and doctors have certain difficulties with technical and physical problems where technicians can be of assistance. The standards developed in this way may be a bit less accurate but they will be much more humanistic. On the contrary, the limits proposed by technicians, based on model studies, even 'millimeter resolution' ones may be accepted as an introductory proposal but nothing more. The approach has nothing in common with real life phenomena, because accuracy of their proposed limits remarkably exceeds the repeatability of biomedical experiments and the accuracy of available field meters. This is well illustrated by the limits shown in Tables 4 and 6, where the limits are given with an accuracy to three or more significant figures! It only confirms a mechanistic, unacceptable character of these proposals. Regardless of the differences between western and eastern standards, one of the drawbacks of existing legal acts, their proposals, recommendations, etc., is either the absence of any metrological character information or these data are excessively detailed [1.10, 1.11]. The latter approach, especially in an international recommendation where any change or modification may sometimes be more difficult than its initial adoption due to long and complex bureaucratic coordination procedures, limits or excludes further improvements of the existing methods or implementation of new methods that could assure much better results (in a technical, operational and interpretational sense) than current methods (instruments). The same may be said regarding calibration methods. It is the author's opinion that any protection standard should include a short statement addressing the metrology. It should contain only the most important parameters and it may be formulated, for instance, in the form: *"the measurement subject is the RMS value of the measured quantity, measured in the extreme conditions within defined frequency ranges. Required accuracy of the measurement is better than ±1 dB when measurements are performed at a distance 5 cm from the nearest radiating source (primary or secondary) or other material media."* The statement contains all the information necessary for correct measurements. Because of the possibility and necessity of measured data comparison and use for statistical, epidemiological and other studies, acceptance of a common measurement methodology is necessary. A statement similar to the one above and included in a protection standard could solve the problem. In this manner, one universal, concise document will give the necessary information for surveying and monitoring services.

Development prospects in this field are almost unlimited. For example, the need to measure fields generated by living organisms, which requires instrumentation far more sensitive than is currently available. These possibilities were introduced in Chapter 9. However, my conclusion still holds that the newest solutions have been designed primarily for research laboratories and specific institutions; whereas the basic equipment for surveying and monitoring services will probably not changed remarkably in the next dozen or so years. We can expect that meters will appear on the market that will allow measurement simplification. However, the measurement technique and methodology is excessively complicated even now and often is not understood even by people with good metrological experience. According to the author's estimation, more than 50 percent of biomedical experiments are performed under unacceptable technical conditions. It is not surprising that metrologists experienced in field measurements use EMF meters equipped with loop antennas for the near-field E-field measurements because *"they are calibrated in the E-field units"*! Another experienced metrologist tried to suggest that the measured field levels generated by a satellite TV receiver well exceeded those permitted by standards when it was actually the secondary field generated in an ungrounded feeder of the roof converter by an FM broadcast station located nearby. Similar tales abound. They confirm that far-field experience in near-field measurements is not enough and that additional training is necessary. The statement can be applied to both the meter manufacturers and to the meter users. Because of the variety of meters available on the market, choosing an appropriate meter may sometimes cause problems, especially for non-technicians. For instance, not every attractively priced meter is worth paying attention to. Table 11 illustrates the wide offering of proposed designs (from selected probes and meters designed and available in Poland) and their metrological possibilities as well as difficulties regarding appropriate meter choice. The table does not imply that these are the best meters in the world, it only shows what is available in Poland to address Polish services, their users' needs and the manufacturers' possibilities. Incidentally, it also illustrates the necessity of matching measuring device parameters to the country's standards.

Based on our discussions, we can try to formulate the basic parameters that characterize a meter's applicability in near-field measurements and what should be taken into account when a meter is evaluated or when a meter is chosen to meet specific requirements:

1. Measured quantity (Chapter 3)
2. Probe sizes (Chapters 4.1, 4.3, 5.1 and 5.4)
3. Frequency response (Chapters 4.2, 5.2 and 5.5)
4. Directional properties of the probe (Chapters 4.4, 5.3 and 7)
5. Measured value: peak, mean or RMS (Chapter 8.2)
6. Dynamic range (Chapter 8.2)
7. Required measurement accuracy
8. Thermal stability (Chapter 8.1)
9. Possible aging effects (mentioned in Chapter 8)
10. Immunity to external EMF (Chapter 8.4)

FINAL COMMENTS

11. Measured field deformations (Chapters 8.4 and 8.5)
12. Complexity of design, operation and service
13. Endurance, reliability, price and cost of exploitation
14. Conformability with national or international standards or specific requirements of the planned measurements
15. Manufacturer's credibility, solidity of a dealer and an accessibility of maintenance
16. Calibration (recalibration) accuracy

The author would like to express his gratitude to Dr. E. Grudzinski for his inspiring comments and the review of the manuscript as well as to Mr. J. Zurawicki for his technical assistance and help preparing the figures. The author would also like to thank the editors for their hard work in adapting his "Penglish" (Polish-English) version of the manuscript into English.

216 ELECTROMAGNETIC FIELD MEASUREMENTS IN THE NEAR FIELD

Meter	Probe	Field	Pattern	Frequency Range	Measuring Range
MEH	AE HP	E	S	0.1 – 30 MHz	0.1 – 10 V/m
	AE 1	E	S	0.1 – 300 MHz	2 – 1,000 V/m
	3AE 1	E	O	0.1 – 300 MHz	2 – 1,000 V/m
	AE 2	E	S	10 – 300 MHz	0.5 – 100 V/m
	AE 21	E	S	0.1 – 600 MHz	1 – 100 V/m
	3AE 2	E	O	10 – 300 MHz	0.5 – 25 V/m
	3AE 2e	E	O	0.1 – 300 MHz	0.5 – 50 V/m
	AE 3	E	S	1 – 100 kHz	5 – 1,000 V/m
	AE 3p	E	S	1 – 100 kHz	5 – 1,000 V/m
	AE 4	E	S	10 – 1,000 Hz	1 – 30 kV/m
	AE 41	E	S	10 – 2,000 Hz	0.1 – 15 kV/m
	AE 43	E	S	50 Hz – 100 kHz	0.2 – 15 kV/m
				1 – 100 kHz	5 – 1,000 V/m
	AH 1	H	S	0.1 – 10 MHz	1 – 250 A/m
	3AH 1	H	O	0.1 – 10 MHz	1 – 250 A/m
	AH 2	H	S	10 – 30 MHz	1 – 250 A/m
	AH 3	H	S	1 – 100 kHz	1 – 250 A/m
	AH 3p	H	S	1 – 100 kHz	1 – 250 A/m
	AH 4	H	S	40 – 1,000 Hz	1 – 500 A/m
	AS 1	S	S	0.3 – 3 GHz	0.1 – 150 W/m^2
	3AS 1	S	O	0.3 – 3 GHz	0.1 – 150 W/m^2
	3AS 3	S	S	0.3 – 40 GHz	0.1 – 150 W/m^2
ME 2	-	E	S	10 – 20,000 Hz	0.01 – 15 kV/m
ME 3	-	E	S	1 – 100 kHz	50 – 1,500 V/m
MH 2	-	H	S	50 – 2,000 Hz	0.2 – 300 A/m
MH 3	-	H	S	1 – 100 kHz	1 – 250 A/m
IIE 1	-	E	S	10 – 45 MHz	7 i 20 V/m
IMR 1	-	S	S		0.025 W/m^2
	-	S			0.1 W/m^2
IMR 3			O	2.45 GHz	2 W/m^2
DPE 1	-	E	O	10 – 300 MHz	100,000 (V/m)^{2}h
DPE 2	-	E	O	50 – 1,000 Hz	100,000 (V/m) h

Indications:
E – electric field,
H – magnetic field,
S – power density,
p – peak value probe.

Directional pattern:
s – sinusoidal,
o – spherical.

Other:
M – meter,
I – indicator,
D – dosimeter.

Table 11. *Selected Polish meters and probes.*

Index

A
Achromatic measurements 150
Amplitude modulation 33
Antenna effect 88, 124

B
Biconical antenna 81

C
Current measurements 41

D
Detectors 62, 181
    diode 62, 159, 167, 191
    dynamic characteristics 167
    electro-optic 185
    Hall-cell 35, 111, 159
    nonlinear 63
    photonic 199
    thermocouple 63, 159
    square-law 72, 139, 159
Dipoles
    elemental electric .. 22, 47, 117
    elemental magnetic 24, 118
    mutually perpendicular
    .... 70, 146
    symmetrical .. 25, 47, 52, 58, 81
Directional pattern 137
    RMS DC summation 149

E/H probe 142
    power density meter 143
    photonic probe 201
    vector summation 147

E
E-field probe 53, 77, 137, 160
    comparison 82
    directional pattern 69, 74
    frequency response 59
    filters 60, 62
    gain 78
    requirements 77
Electric antennas 35
EMF generation 13
EMF meters 216
EMF quantities and units 4
Errors 50
    frequency range 105
    impedance changes 66, 101
    loop antennas 89
    pattern irregularities
    .... 73, 76, 97
    power density 117
    power meters 130
    probe 51
    structure of source 64, 98
    summary of 82, 109
Exposure limits 6, 7, 8

F

Far-field boundary 18
Field averaging 49
Fraunhofer Region 31
Frequency modulation 33
Fresnel Region 31
Future meters 203

H

Hand and lip current 43
Hazard meters 172
High frequency band 56, 93

I

Impedance of free space 17
Impedance of the medium 16

L

Liquid crystals 38
Loop antennas 87
    directional pattern 94
    doubly-loaded 125
    mutual inductance 102
    mutually perpendicular 96
Low frequency band 55, 92
Low pass filter 52, 55, 93

M

Magnetic field measurement 88
Magnetic field probe 87, 138
    directional pattern 94
    equivalent network 91, 106
    frequency reponse
        ....90, 104, 108
Maximal current intensity 44
Maxwell's equations 14, 120
Measurement accuracy
    appropriate usage 213
    detector characteristics 167
    external EMF 172
    field deformations 171
    limitations on 159, 185
    resonant phenomena 175
    thermal stability 159
        correction factors 164
        frequency response 164
    synthesis 210
Medium frequency band
    .... 55, 92, 106
Modulation 33
    AM modulation 169
    CW signals 170
    periodic signals 170

N

Natural EMF environment 2
Near field boundary 18

P

Phase modulation 34
Photonic link 127, 182
    modulators 197
Photonic probe
    description 186
    directional pattern 201
    dynamic characteristics 189

    EMF meter types 187
    equivalent network 190
    frequency response 189
    heterodyne 198
    linearity 199
    sensitivity 195
    structure 190
Polarization 32
    arbitrary 71, 155

circular 137, 148
elliptical 137, 143, 154
linear 137, 148, 153
spherical (ellipsoidal) 154
Power density meters
   frequency 131
   measurment method 120
   operating principle 126
   protection standards 120
Power density measurement
   using antenna effect 124
   by E or H field 116, 121
   measurement error 117
   near-field 119, 135
   versus distance 119
Probe design 153, 190, 203

R
Resonant phenomena 175, 177

S
Selective measurements 32
Specific Absorption 30
Specific Absorption Rate 30, 36
Spectral content 30
Structure of the source 64

T
Temperature rise 36
Thermistor 39
Thermoelement 39
Thermographic measurement ... 40

V
Viscosimeter 40

W
Wideband measurements
    .... 32, 150, 181
Wideband probe 52

Y
Yagi-Uda antenna 175